# SCHOLASTIC ATLAS OF WEATHER

SCHOLASTIC REFERENCE

SCHOLASTIC INC.

New York Toronto London Auckland Sydney
Mexico City New Delhi Hong Kong Buenos Aires

ISBN 0-439-67865-X



12 11 10 9 8 7 6 5 4 3 2 1 4 5 6 7 8 9/0

Printed in the U.S.A. 66

First Scholastic printing, September 2004

*Scholastic Atlas of Weather* was created by:

QA International
329, rue de la Commune Ouest, 3e étage
Montréal (Québec) H2Y 2E1 Canada
**T** 514.499.3000 **F** 514.499.3010
www.qa-international.com

**Editorial Director**
Caroline Fortin

**Editor-in-Chief**
Martine Podesto

**Editor**
Marie-Anne Legault

**Writer**
Donna Vekteris
Marie-Claude Ouellet

**Proofreader**
Veronica Schami

**Graphic Designer**
Josée Noiseux

**Art Director**
Anouk Noël

**Sketch Artist**
Carl Pelletier

**Illustrators**
Anouk Noël
Carl Pelletier

Jocelyn Garner
Jean-Yves Ahern
Alain Lemire

**Photo Researchers**
Kathleen Wind
Gilles Vézina

**Meterologist**
Ève Christian

# Contents

# The ABCs of weather

The different climates and seasons we have on Earth are decided mainly by one thing: where our planet lies and how it moves around the Sun. But there is more to weather than a planet's place in space. Surrounding Earth is a protective layer of air called atmosphere. This atmosphere is like a theater stage, and weather elements—temperature, wind, moisture in the air, and clouds—are like the actors in a play. Together, they put on a spectacular show that is always changing, bringing us weather both foul and fair.

# A place in the Sun

Without the Sun to light and heat it, Earth would be nothing more than a cold, dark, and lifeless rock. Orbiting at a distance of 93 million miles (150 million km) from the Sun, our planet is in an ideal position—between burning hot Venus and freezing cold Mars. With an average temperature of 59°F (15°C), Earth is the only planet known in our solar system that can support life. Because of Earth's shape—a gigantic ball that is slightly flattened out—it does not receive the Sun's heat equally over its surface. Countries near Earth's equator get most of the Sun's rays, and so they have hot climates. Countries lying closer to the polar regions, as well as the North and South poles, get the Sun's rays more indirectly. These regions all have much colder climates.

## Earth's hemispheres and poles

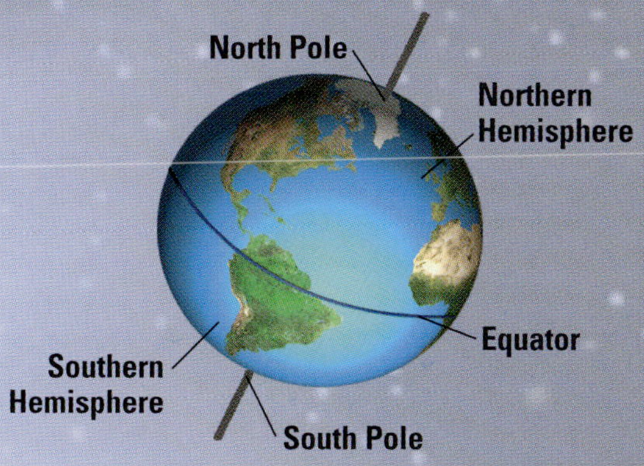


### Superfuel!

The Sun is a giant energy factory. It produces more heat and light in a single second than all the energy produced artificially by humans since the beginning of time!

## The cycle of the seasons

Earth revolves around the Sun on a journey that takes one whole year to complete. It orbits in a slightly tilted position so that the North and South poles are not sitting precisely at the top and bottom of the globe. This means that depending on the time of year, either the Northern or the Southern Hemisphere is receiving more of the Sun's rays. It is the changes in Earth's position that are responsible for the seasons. In July, the Northern Hemisphere is tilted toward the Sun, giving North Americans their summer. At the same time of year, the Australians in the Southern Hemisphere are having their winter. Six months later, the opposite happens.

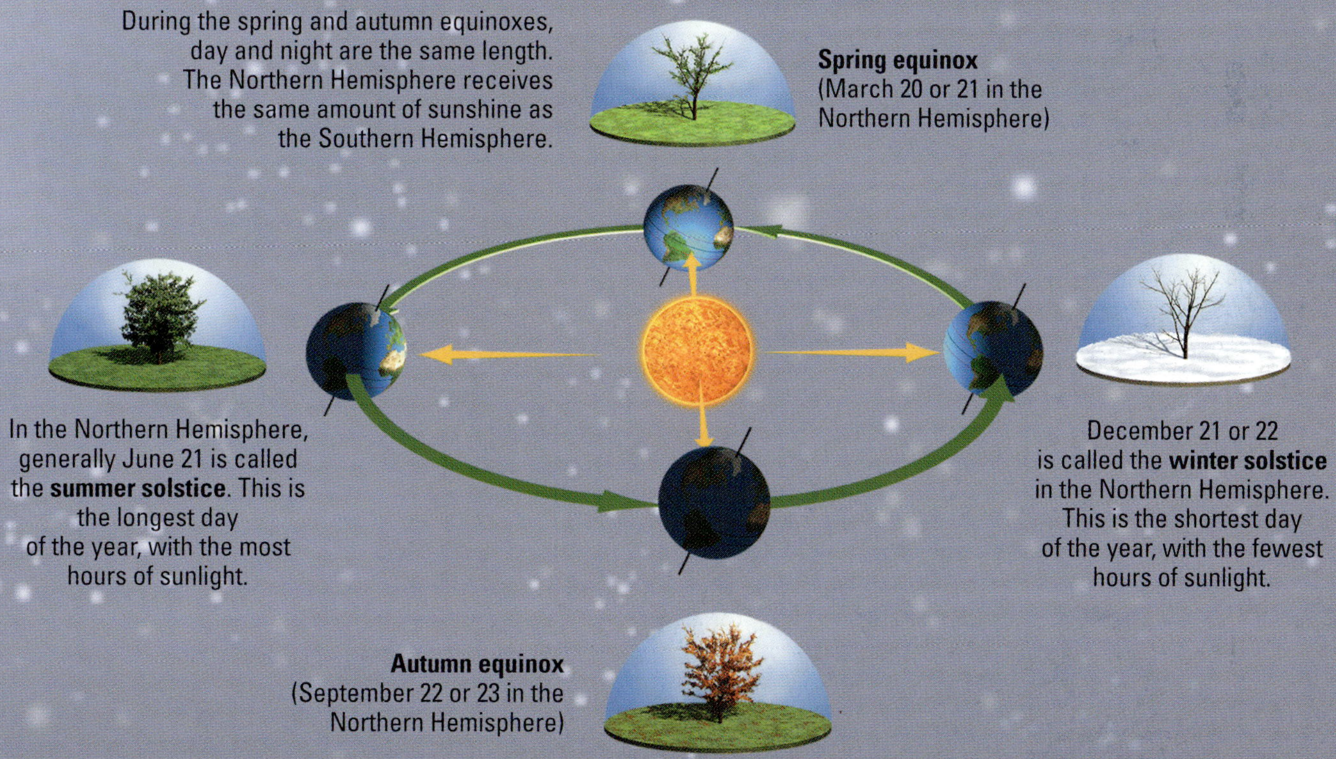

# Earth's trusty shield

The air around us is full of invisible gases. These gases form a protective blanket around Earth that we call atmosphere. All living things on Earth need the atmosphere to survive. It contains the oxygen that allows us to breathe as well as other important gases like nitrogen, carbon dioxide, water vapor, and ozone. Working the way a shield does, the atmosphere helps block the Sun's harmful rays. It also destroys meteorites from out in space that may be heading our way, burning them up before they can hit Earth. Finally, the atmosphere keeps our planet from having extreme temperature changes. Without this protection, our days would be fiery hot and our nights freezing cold!

310 miles (500 km)

50 miles (80 km)

**MESOSPHERE**
(30 to 50 miles [50 to 80 km])
The mesosphere is the coldest layer in Earth's atmosphere. Temperatures here may drop below -148°F (-100°C), and air molecules become rare. Without oxygen tanks, we would suffocate within a few minutes.

30 miles (50 km)

**TROPOSPHERE**
(ground level to 10 miles [15 km])
Most of our weather conditions, like clouds, rain, and storms, are produced in the troposphere. As we move higher in the troposphere, air molecules become fewer and temperatures drop. On the world's highest mountaintops, it is not only very cold, but also hard to breathe because there is little air.

10 miles (15 km)

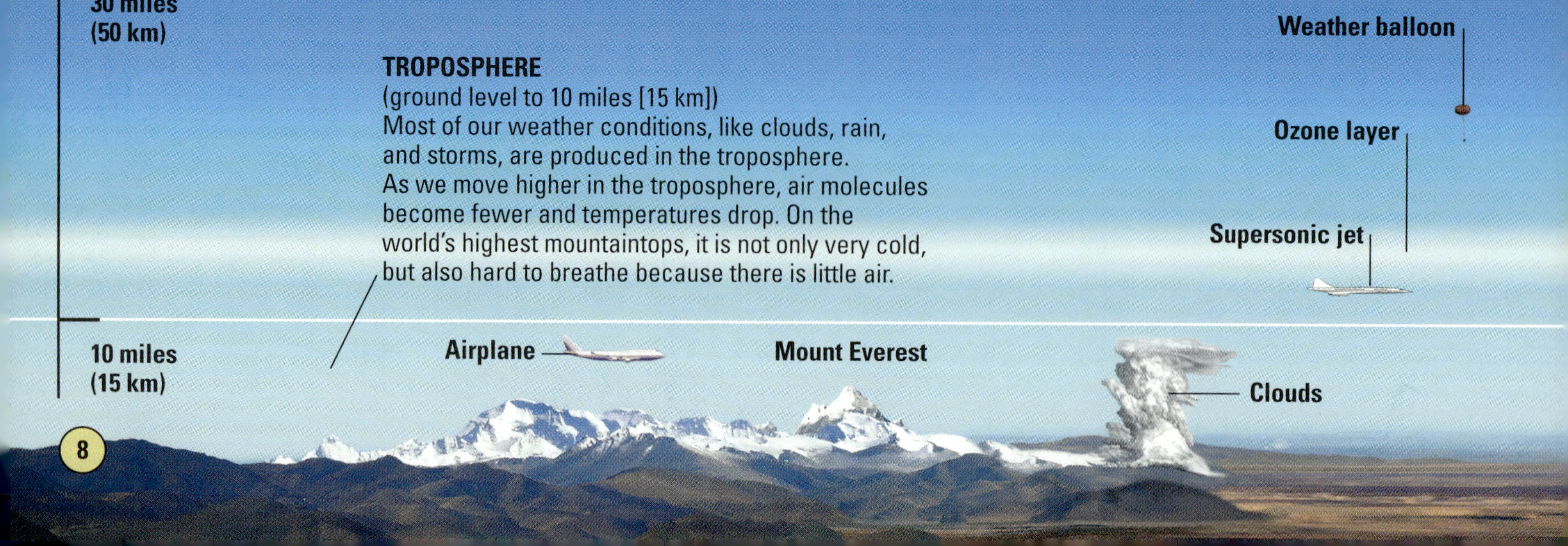

Satellite

**EXOSPHERE**
(310 miles [500 km] and up)
The exosphere is a layer of almost nothing at all. At this height, one air molecule could circle the entire planet before it meets up with another air molecule!

Space shuttle

## As light as air?

Even if we think it weighs nothing at all, air does have weight, which we call atmospheric pressure. Human beings can support hundreds of pounds (kilograms) of air on their shoulders. Without knowing it, we may be holding up as much as one ton of air—equivalent to the weight of a car! We don't feel we are being crushed by this pressure because our bodies have adapted well to our atmosphere.

See activity p. 72

**Meteorites (shooting stars)**

**THERMOSPHERE**
(50 to 310 miles [80 to 500 km])
The thermosphere is the warmest layer in Earth's atmosphere. Because it absorbs most of the Sun's rays, temperatures here may rise above 1,832°F (1,000°C)! Meteorites heading toward Earth are burned up in the thermosphere. Lighting up the night sky, these burning meteorites are called "shooting stars," even though they are not stars at all.

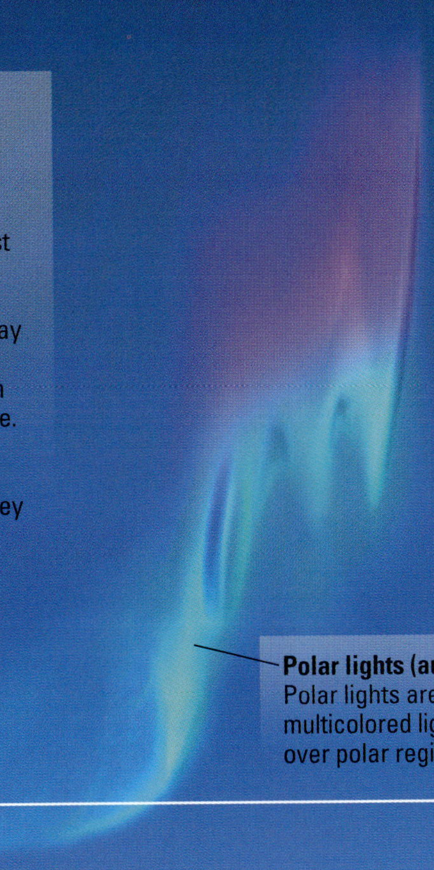

**Polar lights (aurora borealis or aurora australis)**
Polar lights are a spectacular show of shimmering, multicolored light. This phenomenon often occurs in the skies over polar regions.

**STRATOSPHERE**
(10 to 30 miles [15 to 50 km])
Temperatures in the stratosphere actually rise the higher one goes. This is because the stratosphere contains ozone, a gas that produces heat when it absorbs the Sun's ultraviolet rays.

# The hole in the ozone layer

Earth's atmosphere is surrounded by a layer of ozone. Ozone is a gas that absorbs many of the Sun's harmful rays, some of which can cause certain types of cancer. Twenty years ago, scientists noticed that the ozone layer above the Antarctic was shrinking, creating what became known as the "hole" in the ozone layer. The shrinking of this protective shield was caused mainly by CFCs (chlorofluorocarbons), gases once used in refrigerators and aerosol cans. Even though CFCs are no longer produced, the hole in the ozone layer continues to grow. This is because the CFCs already floating in our atmosphere can remain destructive for many years.

# The way the wind blows

Even if we can't see it, we can feel it around us—fluttering in flags, ruffling our hair, and billowing boat sails. Wind is created simply because warm air is lighter than cold air. Moving like a giant Ferris wheel, warm air rises, and its place is taken by cold air close to the ground. As the cold air warms up, it too becomes lighter and starts to rise. In the meantime, the warm air that was above begins to cool down, becoming heavier, and then sinking. It is this large, circular exchange of warm and cold air that produces the wind. The major air currents that travel Earth's surface are called prevailing winds. These winds are constant—generally blowing with the same force and never changing direction. They result from the exchange of large warm and cold air currents between Earth's warm and cold regions.

## Suddenly spring?

A chinook is a warm, dry, local wind that blows downward from the Rocky Mountains toward the North American plains. Chinooks are responsible for sudden and extreme rises in temperature. On January 22, 1943, a chinook wind blew through the Black Hills mountain in South Dakota. The thermometer shot up from -4°F to 45°F (-20°C to 7°C) in just 2 minutes!

# Local winds

Unlike prevailing winds, local winds blow over small areas and often change their direction. The wind that blows on the seacoast is a good example of how a local wind works. During the day, the land warms faster than the sea. The warm air over the land rises, while cooler air over the sea moves inland to take its place. This produces a sea breeze.

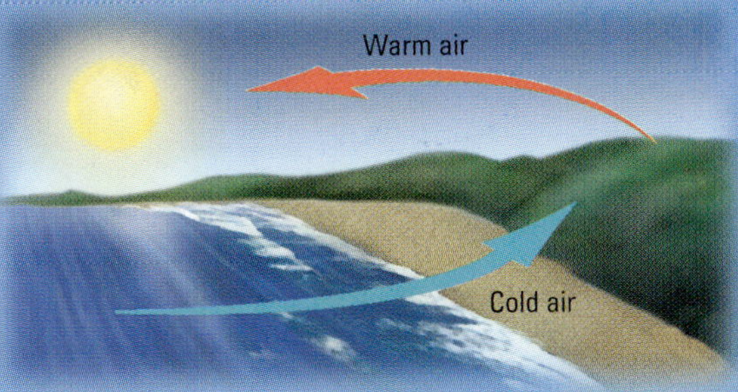


**Sea breeze**

At night, when temperatures are cooler, the opposite takes place. The land cools down more quickly than the sea. As the warmer air over the water rises, its place is taken by cooler air coming from the land. The wind changes direction, and a land breeze is produced.

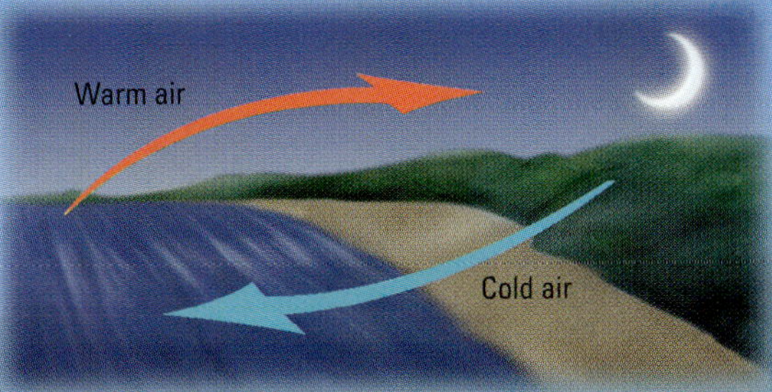


**Land breeze**

# Water and weather

Water is found everywhere—in oceans, lakes, rivers, and even underground. There is also plenty of water in our atmosphere, where it takes the form of an invisible gas called water vapor. Clouds are gigantic reservoirs that store billions of water droplets and ice crystals. They are responsible for much of our weather. Every day, in different parts of the world, the water that clouds hold leaves the atmosphere and falls to Earth in different forms. We call this precipitation. Precipitation is often a delight and a source of wonder to us, but it can also cause a lot of damage and disrupt our daily lives.

**How is a cloud formed?**
Under the heat of the Sun, water from streams, lakes, seas, and the oceans is transformed into vapor and rises in the atmosphere. Higher up, the air is colder, turning the water vapor into tiny droplets that form clouds. The water droplets cluster together and become larger. Once they are too heavy to float in the clouds, they fall to Earth, usually in the form of rain.

See activities p. 74-75

## A mountain of rain

The rainiest place in the world is Mount Waialeale in Hawaii. Every year, this area receives an average of almost 40 ft (12 m) of rain—equal to the height of a four-story building!

# Forms of precipitation

Precipitation depends on two things: the type of cloud the water droplets or ice crystals come from, and the temperatures of the air layers they must pass through on their way to the ground. The main forms of precipitation are shown below.

**Drizzle**
Diameter less than 0.02 in. (0.5 mm)

The clouds that produce drizzle almost touch the ground. The tiny water droplets that fall as drizzle add up to very little water on the ground.

**Rain**
Average diameter: 0.08 in. (2 mm)

Rain usually falls from nimbostratus, thick gray clouds that cover the entire sky. It is called moderate rain when no more than 0.3 inches (7.5 mm) of water collects on the ground in one hour.

**Heavy rain**
Maximum diameter: 0.2 in. (5 mm)

Heavy rain usually falls from large black clouds called cumulonimbus. It is called heavy rain when more than 0.3 inches (7.5 mm) of water collects on the ground in one hour.

**Freezing rain**
Diameter varies

Freezing rain forms when ice crystals encounter a thick layer of warm air that melts them. As the droplets fall on the frozen ground, they freeze again instantly, forming freezing rain.

**Ice pellets**
Diameter less than 0.2 in. (5 mm)

To form ice pellets, water droplets encounter a thin layer of warm air followed by a layer of cold air. The cold air transforms the droplets' walls into ice. Those tiny ice grains are called ice pellets.

**Snow**
Diameter between 0.2 and 1 in. (5–25 mm)

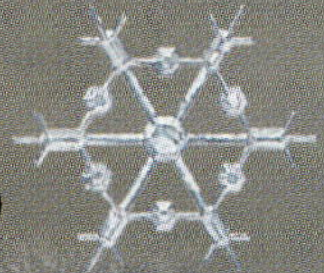

In winter, ice crystals fall to the ground in the form of snow as long as they do not encounter warm air on the way down.

See activity p. 75

# Every cloud tells a story

This one looks like a swan, that one looks like a sheep, and that one over there makes you think of cotton candy! Some clouds look sweet and light, while others look big and threatening. On their own, clouds have the power to make the sky look cheerful or gloomy. It's no accident that clouds come in so many sizes and shapes. Each type of cloud is created by a certain weather condition. By recognizing the different cloud formations, we can better understand the weather that is happening, and can sometimes predict the weather still to come.

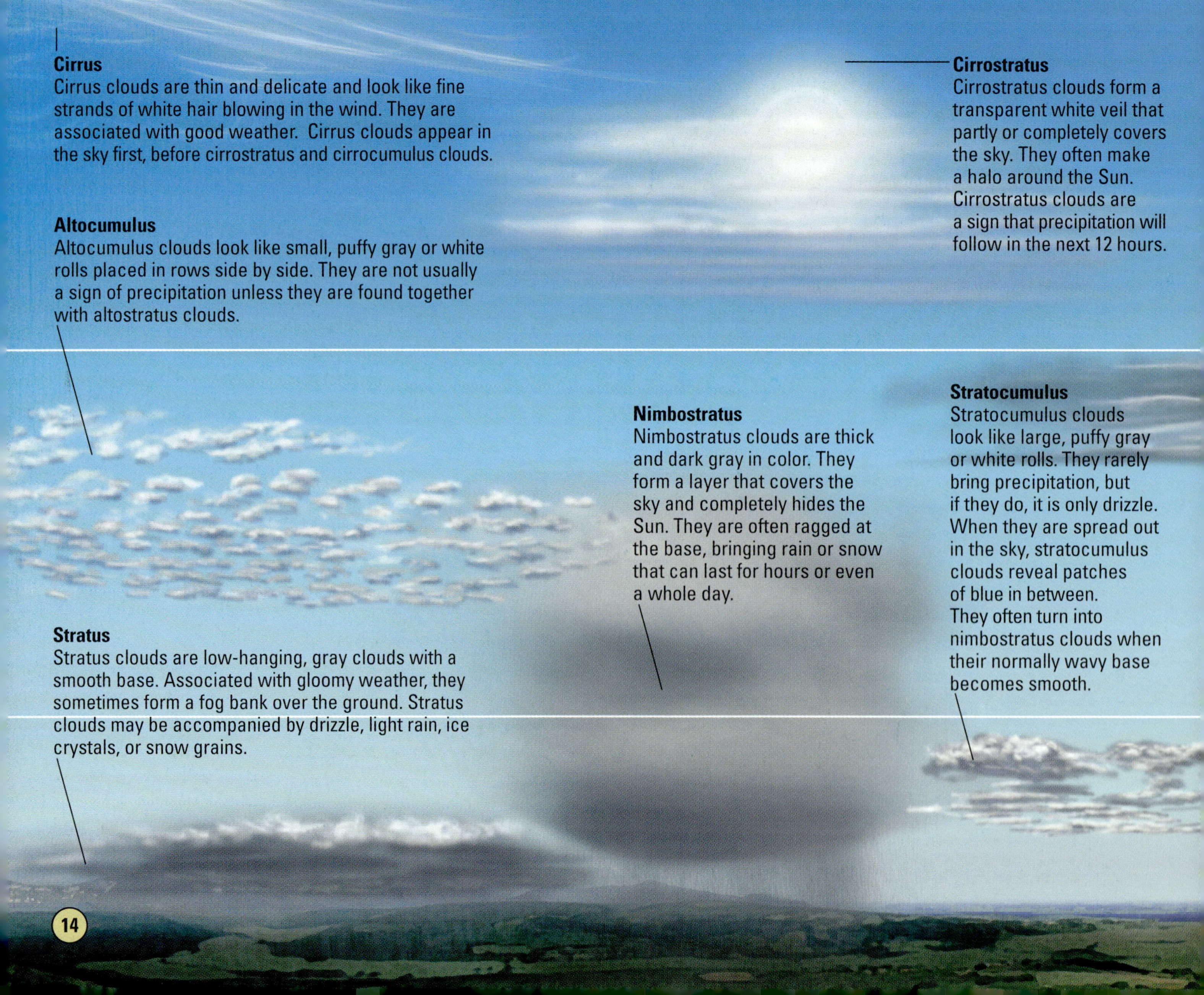

**Cirrus**
Cirrus clouds are thin and delicate and look like fine strands of white hair blowing in the wind. They are associated with good weather. Cirrus clouds appear in the sky first, before cirrostratus and cirrocumulus clouds.

**Cirrostratus**
Cirrostratus clouds form a transparent white veil that partly or completely covers the sky. They often make a halo around the Sun. Cirrostratus clouds are a sign that precipitation will follow in the next 12 hours.

**Altocumulus**
Altocumulus clouds look like small, puffy gray or white rolls placed in rows side by side. They are not usually a sign of precipitation unless they are found together with altostratus clouds.

**Nimbostratus**
Nimbostratus clouds are thick and dark gray in color. They form a layer that covers the sky and completely hides the Sun. They are often ragged at the base, bringing rain or snow that can last for hours or even a whole day.

**Stratocumulus**
Stratocumulus clouds look like large, puffy gray or white rolls. They rarely bring precipitation, but if they do, it is only drizzle. When they are spread out in the sky, stratocumulus clouds reveal patches of blue in between. They often turn into nimbostratus clouds when their normally wavy base becomes smooth.

**Stratus**
Stratus clouds are low-hanging, gray clouds with a smooth base. Associated with gloomy weather, they sometimes form a fog bank over the ground. Stratus clouds may be accompanied by drizzle, light rain, ice crystals, or snow grains.

## THREE LEVELS OF CLOUDS

The highest clouds in the sky (at a level of 3.7 miles [6 km] and up) are cirrus, cirrostratus, and cirrocumulus clouds. At mid-level (between 1.2 and 3.7 miles [2–6 km] high) are the altostratus and altocumulus clouds. The clouds at the lowest level (1.2 miles [2 km] and under) are stratus, nimbostratus, stratocumulus, cumulus, and cumulonimbus clouds. The tops of some of these low-level clouds, like nimbostratus, cumulus, and cumulonimbus, may extend into the higher levels.

**Cumulonimbus**
Cumulonimbus clouds have a dark, scary-looking base and rise very high in the sky. These clouds are like factories for producing thunderstorms, heavy rain or snow showers, hail, violent winds, and tornadoes.

**Cirrocumulus**
Cirrocumulus clouds resemble small, white cotton balls clustered together. They give the sky a wrinkled look. Often found together with cirrostratus clouds, they are a sign that precipitation will follow the next day.

### A roller coaster in a cloud

Cumulonimbus clouds are full of air currents that continually rise and fall. In 1959, during a storm, a pilot in distress parachuted into a cumulonimbus cloud. The violent winds tossed him around in the sky for an hour before he managed to escape the cloud and land.

**Altostratus**
Altostratus clouds form a veil over the sky that is heavier and grayer than that made by cirrostratus clouds. They cover the sky partially or totally, and may allow the Sun to show through. Altostratus clouds are a sign that rain showers will follow shortly.

**Cumulus**
Cumulus clouds are pretty and white and look like cotton. Scattered across a blue sky, they are a sign of good weather. If they grow taller on top, however, they become cumulonimbus clouds and bring precipitation that will last one or two hours.

# Ups and downs

The Sun is the engine that produces Earth's temperatures. Depending on our distance from the equator and the time of year, we receive more or less of the Sun's heat. Other factors besides the Sun affect temperatures on Earth. For example, air in the mountains is usually cooler than air at sea level. Coastal areas often have milder climates year-round. Inland on continents, however, there are much bigger differences between summer and winter temperatures. Wind and humidity—or moisture in the air—greatly affect our sense of heat and cold, no matter what the thermometer tells us. Usually, the harder the wind blows, the colder the temperature feels. Humidity, on the other hand, can make the temperature seem hotter or colder depending on the time of year.  It makes us feel like we are boiling in the summer, but freezing in the winter!

## As night fell . . .

Normally, temperatures rise during the day, while the Sun's rays are warming the earth. After sunset, temperatures fall. Sometimes this drop in temperature can be quite sudden and serious. During the night of January 23–24, 1916, in Montana (U.S.A.), the temperature plunged from 45°F (7°C) to -56°F (-49°C). That is a difference of 101°F (or 56°C)!

# The three temperature scales

The thermometer, as we know it today, was invented in the early 1700s by Daniel G. Fahrenheit, who lent his name to this system. Zero degrees Fahrenheit was the coldest temperature that had ever been recorded up to that time. The Fahrenheit scale is still used today, mainly in the United States.

Most countries, however, have adopted the Celsius system of measurement, developed by Anders Celsius in 1742. On the Celsius scale, zero degrees corresponds to water's freezing point, while 100 degrees represents its boiling point.

Scientists prefer William Thomson Kelvin's method. Zero degrees Kelvin (-273°C/-460°F), also called absolute zero, is the coldest temperature that scientists believe can be reached.

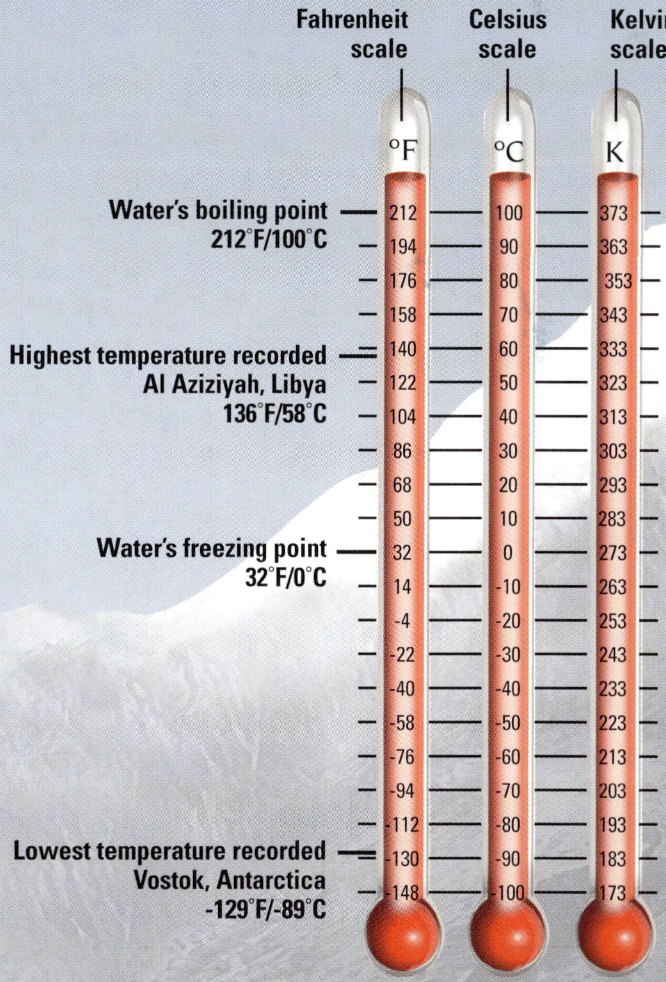

# When weather runs wild

**Raging winds whip up a tornado. The tornado rips through the countryside, tearing down everything in its path. A sudden downpour fills a river. The river overflows its banks and the rushing waters flood the town. A hailstorm hits a farm, raining chunks of ice that crush the summer's crop. Extraordinary weather events have the power to disrupt daily routine, create disorder and chaos, and even bring entire populations to a standstill.**

# Destructive whirlwinds

They shatter windows, uproot trees, and tear the roofs off houses. They can send cars and locomotives—as well as animals and people—through the air! Tornadoes are the most dangerous meteorological phenomenon on Earth. With their gigantic funnels, spiraling down from cumulonimbus clouds, tornadoes suck up everything in their paths. Their violent, spinning winds sound like the roar of a jet plane. Accompanied by thunderstorms, rain, and often hail, tornadoes can sweep across a landscape in just minutes, causing terrible damage for miles around.

## Tornado tragedy

The most destructive tornado of all time was the Tri-State Tornado. On March 18, 1925, it traveled more than 200 miles (320 km) across the United States, passing through Missouri, Illinois, and Indiana. The Tri-State Tornado took 695 lives.

# How are tornadoes formed?

Luckily, not all cumulonimbus clouds produce tornadoes! All the right weather conditions must be present to create these violent storms.

1. Rapidly moving cold air at a high altitude crosses over slow-moving warm air nearer to the ground. The two air currents start to swirl, one over the other, until they form a giant tube inside the cumulonimbus cloud.
2. Warm ground air rises up to the cloud and pushes on the roll of swirling air, tipping it over to an upright position.
3. The funnel shape that is formed at the base of the cloud creates a vacuum that sucks up warm ground air at an increasingly faster rate. The funnel stretches downward, growing longer until it reaches the ground and becomes a tornado.

# Spinning out of control

Tornadoes are difficult to predict and impossible to stop. Because warnings often come just minutes before the tornadoes themselves pass through an area, not many people are prepared to deal with them. The basement of a house, a small bathroom, or a closet can provide shelter when a tornado strikes. If caught outdoors, a person can take cover in a ditch or in any slight depression in the ground.

The part of the world most affected by tornadoes is the United States. Almost 1,000 twisters strike the U.S.A. every year! The great American plains, which are found in Texas, Oklahoma, Kansas, and Nebraska, create the best conditions for tornadoes to form. This area is where hot, humid air from the Gulf of Mexico in the south meets up with cool northern air from Canada.

# The Fujita scale

Tornadoes may cause destruction and even death as they pass through an area. Fortunately, all tornadoes don't have the same amount of power. While one may knock down a house, another may do nothing more than bend TV antennas! Tornadoes are rated according to how much damage they cause. This way of classifying tornadoes is called the Fujita scale, named after tornado expert Theodore Fujita.

**F-0** **40 to 72 mph (64 to 116 km/h)**
TV antennas bent, chimneys and road signs damaged, tree branches broken

**F-1** **73 to 112 mph (117 to 180 km/h)**
Roof shingles torn off houses, windows broken, mobile homes overturned, small trees uprooted

**F-2** **113 to 157 mph (181 to 252 km/h)**
Mobile homes demolished, large trees uprooted, small vehicles shifted, wooden buildings destroyed

**F-3** **158 to 206 mph (253 to 330 km/h)**
Roofs and some walls of houses collapsed, heavy vehicles overturned

**F-4** **207 to 260 mph (331 to 417 km/h)**
Houses demolished; cars, trucks, and train cars lifted into the air; objects weighing several hundred pounds blown away

**F-5** **261 to 318 mph (418 to 509 km/h)**
Houses ripped from their foundations and carried considerable distances, reinforced concrete buildings damaged, cars thrown through the air

## A deep sleep

In 1981, a tornado lifted a baby out of his stroller and sent him flying 50 ft (15 m) up into the air. The whirlwind then managed to gently set the baby back down on the ground, some 300 ft (90 m) away from the stroller. All this happened without waking the baby!

# Mystery mists

Swirling white blankets of mist and fog cover the land in mysterious beauty. Created on the earth's surface, mist and fog are actually clouds that hover close to ground or water. Like clouds, mist and fog are made up of small water droplets that float in the air. On the sea, on lakes, and on rivers, large banks of fog are a real danger to both professional and amateur sailors. At ground level, fog reduces visibility on roads and bridges, and can make car travel dangerous as well as difficult.

**Condensation**
Air contains water in the form of an invisible gas called water vapor. The amount of water vapor in the air changes with the temperature. Warm air can contain more water vapor than cold air. When air cools down and can no longer hold any more vapor, we say that it has reached the dew point. At this precise moment, the vapor is transformed into visible water droplets that can be dew, mist, or fog. This process is called condensation.

Mist

Fog

**Clouds of mist and fog**

Mist and fog are created as water vapor condenses in the air. The cloud created by fog may reduce visibility to anywhere from several feet to half a mile (1 km). When the water droplets that make up the fog are more widely spread apart, the fog is called mist. In mist, visibility is reduced to anywhere from half a mile to 3 miles (1 to 5 km).

## Fogbound

The areas bordering oceans are among the most affected by mist and fog. With more than 2,550 hours of fog each year, Cape Disappointment, bordering the Pacific Ocean in Washington state, is the foggiest region in the United States.

**Dew**

Dew is formed when air close to the ground cools down to the dew point and condenses. This mostly occurs at the end of a calm and clear night. When it comes into contact with cold grass or other objects on the ground, the water vapor in the air condenses and forms sparkling dewdrops.

# When rain rules

A heavy rain falls on the town and the surrounding countryside. The sudden downpour is too much for the ground and the nearby river to soak up in a short amount of time. The river starts to rise, and then overflows its banks. Confusion and destruction follow as the town begins to flood. During severe floods, furniture, trees, rocks, and even vehicles are carried away by the rushing waters. Drivers are caught inside their cars, houses fill with mud, and the town's drinking water becomes too dirty for people to use. Every year, more damage is caused and more lives are lost because of floods than because of lightning, hurricanes, or tornadoes.

**The ground collapses**
Very heavy rains falling in the mountains often cause landslides. Under the pressure from all the water hitting the ground, tons of mud and rocks may fall down the mountainsides with amazing speed. Landslides can bury roads under debris, uproot trees, and destroy houses.

**Levee**
A levee is a barrier built alongside a waterway. Its purpose is to protect the land from overflowing water.

## The power to push a car

It takes just 2 ft (60 cm) of water to make a car float. If floodwaters are flowing quickly, they can create a disaster! A powerful water current can lift heavy objects, such as cars, and move them long distances.

# A spectacle of light and sound

Thunderstorms are terrifying and yet beautiful to watch. Featuring lightning as well as thunder, they are one of nature's most common weather phenomena. In fact, the earth is struck by lightning almost 100 times per second! Thunderstorms are created in the giant cumulonimbus clouds found in the sky on hot summer days. The show they put on is hard to top! The amount of electrical energy produced in a storm can set houses and forests on fire, cause electrical blackouts, and damage airplanes. Traveling through electrical wiring, lightning can also damage computers. Very humid tropical regions are most often hit by thunderstorms. The Indonesian city of Bogor, in the South Pacific, holds the record. In 1916, storms occurred for 322 days, that is, about six days a week!

## Electrifying!

A single bolt of lightning may produce enough current to run 8,000 toasters all at the same time!

**What is thunder?**
When lightning crosses the sky, it makes the air around it so hot that it produces a powerful explosion. This sudden movement of air around the lightning bolt creates a loud sound: thunder!

# How lightning is formed

The winds found in a cumulonimbus cloud create a disturbance inside it. Rain droplets and ice crystals in the cloud begin to rub together and crash into one another, creating tiny electrical particles. These particles, charged with either a positive or a negative current, are attracted to one another. This attraction between opposite charges causes the particles to move toward one another, creating lightning. The following illustration shows how lightning moves between a cloud and the ground.

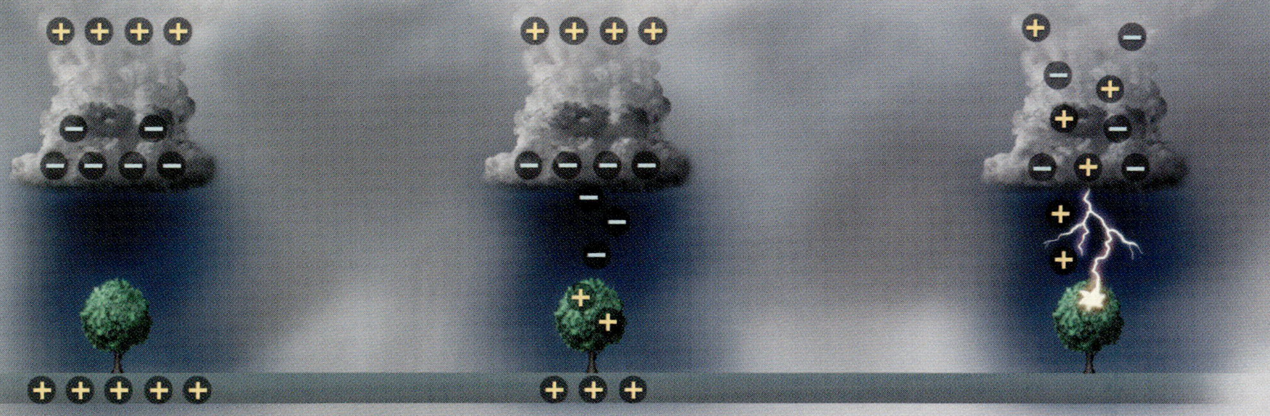

1. Positive charges collect near the top of the thundercloud, while negative charges collect near the base. Positive charges gather on the ground below the cloud.
2. The negative charges in the cloud are attracted to the positive charges on the ground. As they move toward one another, they create invisible sparks.
3. When the two invisible sparks meet, they create a path that makes the positive charges travel from the ground up to the cloud. This upward movement creates a visible lightning bolt.

**When lightning strikes!**

When a thunderstorm hits, it is important to take shelter immediately, either in a building or in a vehicle with closed windows. The power of the electricity released by the lightning is strong enough to cause severe injuries, even death. In spite of the danger, it's surprising to know that four out of every five people struck by lightning actually survive!

# Buried landscapes

Billions of people around the world have never seen snow. Yet some cities in the Northern Hemisphere, like Montreal, Chicago, and Detroit, each receive millions of tons of snow every winter! When a lot of snow falls in a short period of time, it is called a snowstorm. Snowstorms close roads and airports, making it difficult or impossible to travel. If the snow is heavy enough, it may damage houses, electrical lines, and trees. For a big snowstorm to be produced, the right type of clouds must form. This requires temperatures close to freezing, combined with wind and a lot of water vapor in the air. If a snowstorm continues for three hours or longer with wind blowing at a speed of 35 mph (56 km/h) or more, and visibility is reduced to 1/4 mile (400 m) or less, the snowstorm is called a blizzard.

## Snowed under!

The biggest amount of snow to ever fall in a single storm occurred at Mount Shasta Ski Bowl, California, in 1959. Between February 13 and 19, 15 ft 9 in. (4.8 m) of snow was dumped on the area. That's almost the height of the average one-story house!

## Temperature and the shape of snowflakes

**Icy jewels**

There are no two snowflakes alike. Each one is made up of thousands of crystals of ice that have joined together in unique ways. Snowflakes come in two basic shapes: stars and needles. The temperature and amount of water vapor in the air decide what form these icy jewels will take. As different as they may seem, however, all snowflakes have something in common: six sides. This particular shape comes from the way the water freezes into ice crystals.

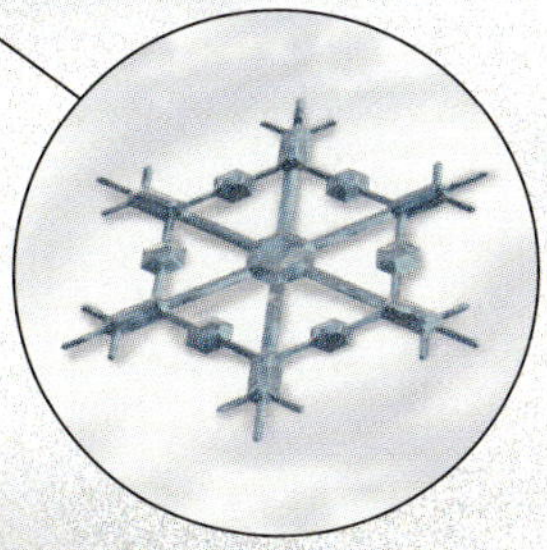

# A world on ice

Under the magical spell of ice, the countryside sparkles like a fairytale scene frozen in time. In spite of its beauty, however, freezing rain is dangerous. It can make streets and sidewalks as slippery as skating rinks. When it lasts for several days, an ice storm becomes extremely destructive! It can snap electrical cables and topple the poles that hold them, leaving people without light and heat for their homes. In the forest, the heavy ice is capable of splitting apart thousands of trees. Wild animals might have trouble finding food because the plants they normally feed on during the wintertime may be trapped under a thick layer of ice.

## The Great Ice Storm of '98

An ice storm hit eastern Canada, New England, and upstate New York in January 1998, when almost 4 in. (10 cm) of freezing rain fell in just six days! When electrical towers and poles collapsed, 4 million people were plunged into darkness and extreme cold. In some regions, the blackout lasted five weeks!

## How does freezing rain form?

For freezing rain to form, all the right conditions must be present. The air close to the ground must be cold, below the freezing point (32°F/0°C). The layer of air close to the ground must be topped by warm air (above 32°F/0°C), and then topped by yet another layer of cold air. This is how freezing rain is formed:

1. The snow that falls from the clouds meets warm air and is changed into rain.
2. As these raindrops fall through the layer of colder air near the ground, they chill but do not freeze entirely, even if they are colder than the freezing point. At that moment, the raindrops are supercooled.
3. When they finally hit the ground or come into contact with objects that are colder than 32°F (0°C), the supercooled raindrops freeze almost instantly and form a thin coating of ice.

# The sky is falling

A hailstorm hits without warning. It batters the roofs of houses and shatters windows, dents cars, and destroys entire crops. With beads of ice pounding the ground at more than 60 mph (100 km/h), hail is probably the most destructive precipitation on Earth. Hailstorms rarely occur in hot countries because the hailstones melt long before they reach the ground. Hailstorms happen mainly in the spring and summer months in temperate zones, which are areas with four different seasons. They develop during thunderstorms and very humid weather, when winds are strong enough to hold up the hailstones that form in the clouds. The central region of North America is the area most often hit by hailstorms. In Colorado in 1984, a storm left the people of Denver with hailstones up to their knees!

## Dangerous hailstones

Hailstones are usually about the size of peas, but they may be as large as grapefruits. At this size, they become weapons that can seriously injure humans and animals. The biggest hailstone ever found in the United States was the size of a honeydew melon!

# How are hailstones formed?

Hailstones are formed inside cumulonimbus clouds, where freezing water droplets are put in motion by strong winds. This is how a hailstone is formed:

1. Moved by downward winds, a drop of freezing water goes to the lowest and warmest part of the cloud. There, it picks up a layer of clear ice made up of the surrounding water droplets and starts to form a hailstone.

2. Lifted by upward-moving winds, the tiny hailstone then rises to the top of the cloud. There, freezing air causes the water droplets that are clinging to the hailstone to freeze instantly. It grows in size as it is covered in a coating of white ice.

3. After moving up and down several times by the winds in the cloud, the hailstone is gradually coated in more layers of ice. Once it grows too heavy to be supported by the winds, the hailstone falls to the ground.

**Layers like an onion**
Hailstones move up and down inside cumulonimbus clouds for 5 to 10 minutes before falling to the ground. While traveling inside the cloud, they may pick up as many as 25 layers of ice.

# From one extreme to another

Climate is different from one region to another. It is affected by temperature, winds, and the amount of rain or snow an area receives. One extreme is found in Earth's 18 million square miles (48 million sq km) of hot desert. Here, temperatures often reach more than 122°F (50°C), and violent winds whip up the blazing sands. The other extreme is found at the North and South poles. These areas receive very little direct sunlight and have very cold, dry climates. The North and South poles have the worst blizzards on Earth, with raging winds that tear along the open landscape at more than 185 mph (300 km/h).

**People of the desert**
People who survive in the desert often live in oases. These are areas in which some water, trees, and plants may be found. Men, women, and children who live in the desert usually wear long, loose robes. Their clothing protects them from the burning rays of the Sun and allows their perspiration to escape easily. A desert house has thick walls to protect the people indoors from extreme temperature changes outside. Because its walls take a long time to cool down and to heat up, the inside of the house stays cool during hot days and remains warm during cold nights.

**People of the far North**

Although very few human beings can survive in Earth's cold deserts, the Inuit of the Arctic manage to live there year-round. Staying warm in this unfriendly climate requires them to wear layers of thick fur clothing. The diet of the Inuit is made up of mainly meat and animal oils. This type of food gives the Inuit the energy they need to fight the cold.

See activity p. 71

## Never enough rain

Although the humid tropical rain forests of South America receive as much as 10 ft (3 m) of rain each year, some deserts may remain completely dry for years at a time. In 1971, the Atacama Desert in Chile received its first rainfall in 400 years! This area has the driest climate in the world.

**Record-breaking temperatures**

The Vostok region, located in the middle of the Antarctic ice cap, is the coldest place on Earth. On July 21, 1983, this area recorded a temperature of -128°F (-89°C)!

The city of Al Aziziyah in Libya holds the record for the highest temperature on the planet. On September 13, 1922, the city recorded a temperature of 136°F (58°C)!

## Dusting up a storm

China is one of the countries hardest hit by dust storms. These storms travel long distances, not only within China but to other countries as well. The clouds of dust may be blown halfway around the world, increasing air pollution in places as far away as the United States!

**How is a dust storm created?**
The blazing Sun heats the ground to a high temperature. The air just above the ground also heats up, and then begins to rise. This air movement sometimes creates strong winds capable of lifting large amounts of sand or dust off the ground. A cloud of particles begins to form just above the ground. In time, the cloud becomes larger and thicker, making it difficult to see.

# Scattered to the winds

Raging winds stir up an enormous, dense cloud made of grains of sand, rising 10,000 ft (3,000 m) in the air. This cloud may travel for days, journeying thousands of miles and even crossing the ocean! Although they are usually created in the desert, sandstorms may also occur in regions where the ground has turned to dust. This can happen if it hasn't rained in a very long time, and if the soil has been worn out by too much farming or raising of animals. A storm may darken the sky so much that you need to hold a lamp right in front of your face to be able to see. The tiny particles work their way into houses, clothing, and even food. Blown by powerful winds, they also sting the skin; go into the eyes, nose, and mouth; and can even cause permanent damage to the lungs.

**A dust "tornado"**
Dust devils are whirlwinds that often form in very dry regions like Australia, the Middle East, and the southwestern United States. Although they resemble miniature tornadoes, their effects are much less damaging. These whirlwinds rise usually no higher than about 1,000 ft (300 m) in the air and generally last only a few minutes.

# Setting the world on fire

Every year, hundreds of millions of acres (hectares) of forests around the world go up in smoke. Australia, California, and France's southern coast are the areas most affected by forest fires. These regions have hot, dry, windy climates and vegetation that burns easily. Most fires in forests, grasslands, and bush areas happen because of human carelessness. Sometimes these disasters strike without anyone's help at all. Thunderstorms are mainly responsible for natural forest fires. The sparks created by lightning strikes are often enough to set fire to vegetation. Weather conditions during a forest fire have a lot to do with how much destruction will occur. If the weather is hot and dry and the winds are strong, the damage may be widespread. Sometimes the only thing to bring the raging flames under control is the rain.

## Adding fuel to the fire

Eucalyptus trees, which are native to Australia, have a lot of oil inside them. When the trees burn, this oil becomes a dangerous fuel that causes the trees to explode, spreading the flames quickly. If the weather is windy, the fires may eat up an area of 1.5 square miles (4 sq km) in just 30 minutes! This is equal in size to 800 football fields!

**Cooling down or warming up?**

Forest fires affect climate. They give off gases and tiny particles that remain suspended in Earth's atmosphere for a long time. These particles increase air pollution and block the Sun's rays, causing a slight cooling in the region affected by the fire. On the other hand, the fires give off a lot of carbon dioxide. Called a "greenhouse" gas, carbon dioxide traps heat in Earth's atmosphere and contributes to overall global warming.

Fires can turn forests into wastelands. Amazingly, these disasters can be helpful for nature. Fires remove trees that are diseased or too old and allow new and healthy vegetation to grow.

# Monsters from the sea

The people who live in North America and the Caribbean call them "hurricanes." The inhabitants of southeast Asia name them "typhoons." In Australia, they are called "willy-willies," and in the Indian Ocean, they are known as "cyclones." No matter what name they may go by, these storms are a nightmare for everyone who lives in a tropical region. Hurricanes can spread out over wide areas of ocean, sometimes hitting coastlines. With their clouds rolling in giant whirlwind formations, they bring with them extremely violent winds and heavy rains. In just seven to nine days, these giants of the Tropics may travel thousands of miles (kilometers), threatening boats as well as the people living in coastal regions.

**The eye of the hurricane**
The eye is a calm zone inside the hurricane. The winds in this zone are weak, the sky is often clear, and there is almost no rain. The diameter of the eye can vary a lot, but the average size is 18 miles (30 km).

**Betsy, Carlos, or Danielle?**
Since 1979, meteorologists from the World Meteorological Organization (WMO) have been naming hurricanes. Each year, they come up with an alphabetical list that alternates between boys' and girls' names in English, Spanish, and French. Every time a hurricane is born, it is given a name from the official list. The first hurricane of the year will have a name starting with the letter "A," the second hurricane with the letter "B," and so forth.

## A powerful engine

While it condenses, water vapor releases a lot of energy. In a single day, a hurricane gives off enough energy to supply the electrical needs of the United States for six months!

# The formation of a hurricane

With the help of the Sun, warm and humid air starts to rise above the ocean's surface. A group of storm clouds begins to form. Inside the center of the biggest storm cloud, an invisible "chimney" develops. Air is sucked in at the bottom of the chimney and rises in a spiral pattern. As the air nears the top, it condenses, forming clouds that roll in a whirlwind pattern. This movement of rising air near the center of the storm cloud acts as a giant vacuum cleaner. As long as the warmth of the ocean feeds it, the hurricane's chimney continues to suck in air.

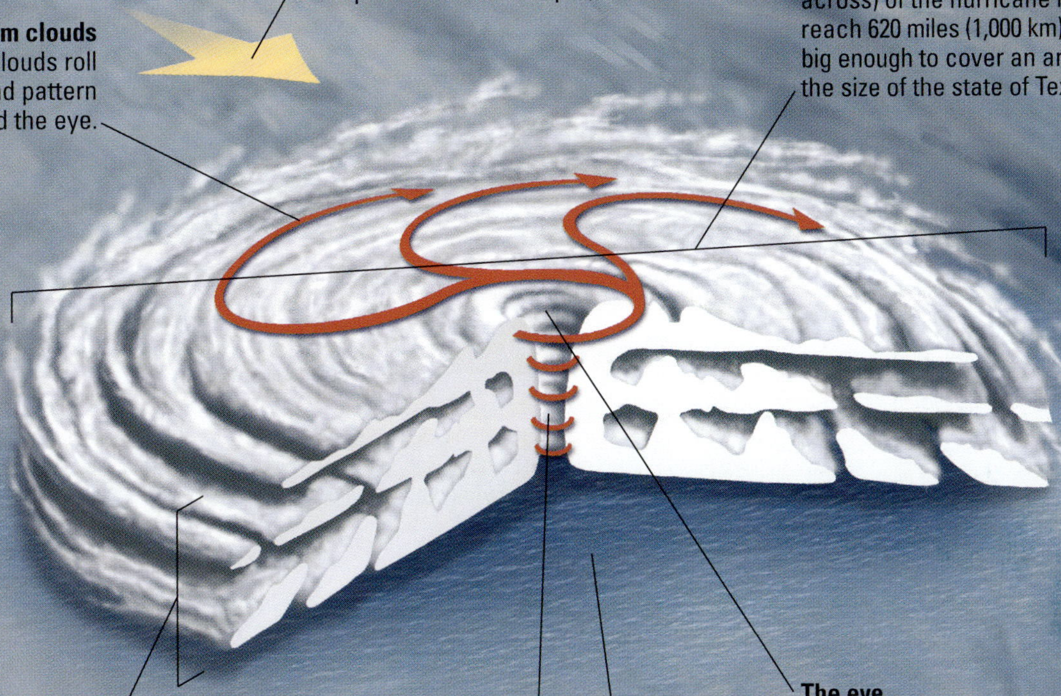

# The dangers of hurricanes

It is impossible to predict several days ahead exactly what path a hurricane will take. It may suddenly change direction or go back in the direction from which it came. If it is traveling over cold water currents, it may even die out. Sometimes these wind and rain monsters manage to make their way to the shores of a continent or an island, with terrible results. The storm's heavy rains cause landslides and make rivers overflow. Its screaming winds uproot trees and tear roofs off houses. The winds also create gigantic waves on the ocean, which crash against the shore with a terrible sound that can be heard for miles (kilometers) around. Luckily, hurricanes die out quickly once they touch down on land. Left without the warm water of the ocean that feeds them, the winds lose their strength, even if the rain continues to fall for several days afterward.

## Are hurricanes useful?

Hurricanes are so destructive that it is hard to believe that they can do any good. Yet they do play an important part in supplying much-needed rain to hot countries that have been affected by drought. Many regions actually owe the success of their harvests to hurricanes!

# The Saffir-Simpson scale

Since the 1970s, the National Hurricane Center in the United States has been rating hurricanes according to their different features, including wind speed and tide height. The Saffir-Simpson scale helps scientists estimate how dangerous a storm may be and how much damage it may do.

| | 1 | 2 | 3 | 4 | 5 |
|---|---|---|---|---|---|
| **Wind speed** | 73 to 94 mph<br>118 to 152 km/h | 95 to 109 mph<br>153 to 176 km/h | 110 to 129 mph<br>177 to 208 km/h | 130 to 155 mph<br>209 to 248 km/h | More than 155 mph<br>More than 248 km/h |
| **Tide height *** | 4 to 5 ft<br>1.2 to 1.7 m | 6 to 8 ft<br>1.8 to 2.6 m | 9 to 12 ft<br>2.7 to 3.8 m | 13 to 18 ft<br>3.9 to 5.5 m | More than 18 ft<br>More than 5.5 m |
| | Trees and shrubs damaged; trailer homes, docks, and moorings of small boats damaged | Small trees uprooted; trailer homes seriously damaged; some roofs damaged | Leaves ripped off trees; large trees uprooted; trailer homes destroyed; some roofs, windows, and doors of houses damaged | Traffic signs thrown to the ground; roofs, windows, and doors of houses seriously damaged | Some buildings destroyed; many roofs of houses damaged |

*** Above normal**

**What are storm surges?**
Hurricanes have a vacuum effect on the ocean. The water is pulled toward the hurricane, causing it to "pile up" like a small mountain. Once it reaches the coast, the water spills onto the land, suddenly flooding large areas. In 1970, in Bangladesh, a storm surge raised the level of the sea by 39 ft (12 m). This is equal in height to a four-story building! As a result, 300,000 lives were lost.

# A planet under many influences

**Earth's climate is influenced by its own land formations and water movement. Mountains, as well as traveling ocean currents, have an effect on weather patterns in different parts of the world. Climate is also affected by rare phenomena. Events such as a major volcanic eruption or the collision between Earth and an asteroid in space may be responsible for changing the world's climate in dramatic ways.**

# Spectacular explosions

Every year, approximately 60 volcanoes on Earth become active. Some release large amounts of gas, ashes, and dust into the atmosphere. Blown by winds, these particles may float around in the atmosphere for months, even years. Forming clouds, the volcanic material stops some of the Sun's rays from reaching Earth's surface. This can cause a slight drop in temperatures worldwide. Another phenomenon causes temperatures around the world to drop on a much larger scale—the collision of a comet with our planet. When a comet hit Earth some 65 million years ago, that collision produced enormous clouds of dust that blocked the Sun and dramatically cooled Earth's climate. Today, astronomers keep a constant watch for comets that could be heading our way. Most astronomers think that another collision of this kind is not likely to happen.

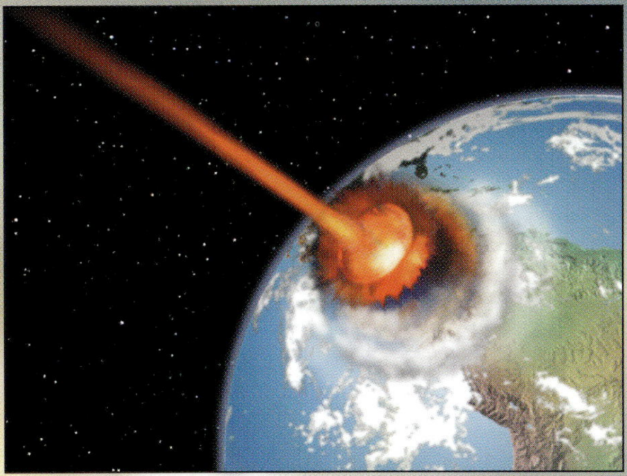

**The end of the dinosaurs**
Scientists believe that the comet that hit Earth 65 million years ago was responsible for the disappearance of many different species of animals, including dinosaurs. After the comet struck, the gigantic dust clouds that covered our planet blocked the Sun's rays for several months. Without the heat and light of the Sun, many plants died. Deprived of their food, the plant-eating dinosaurs disappeared, which in turn led to the starvation of the meat-eating dinosaurs.

## "The year without a summer"

In 1815 in Indonesia, in the South Pacific, the volcano Tambora erupted. It was the largest volcanic eruption ever recorded. The following year, areas as distant as Europe and North America reported temperatures far below normal. For that reason, the year 1816 became known as "the year without a summer." This cooling down of Earth's climate had a negative effect on harvests and caused food shortages in some countries.

# Shaping Earth's climate

Two-thirds of Earth's surface is covered with water. The remaining one-third is made up of land. The combination of oceans and landmasses not only makes up Earth's surface, but also influences Earth's climate. The large currents of warm and cold water that circulate in the world's oceans have an effect on the weather of nearby land. The Gulf Stream, for example, is a warm ocean current that flows by the coast of England. It helps give that country particularly mild winters. Mountains also have a direct effect on local weather. They influence the amount of rain or snow that falls in the surrounding areas.

# How mountains affect precipitation

When air containing water vapor (moist air) encounters a mountain, it is forced upward. As it rises, the air cools and condenses to form clouds near the mountaintop. This phenomenon is responsible for heavy rainfall or snowfall on the side and at the top of the mountain, where the clouds are formed (mountainside facing the wind). The other side of the mountain receives very little precipitation (mountainside facing away from the wind).

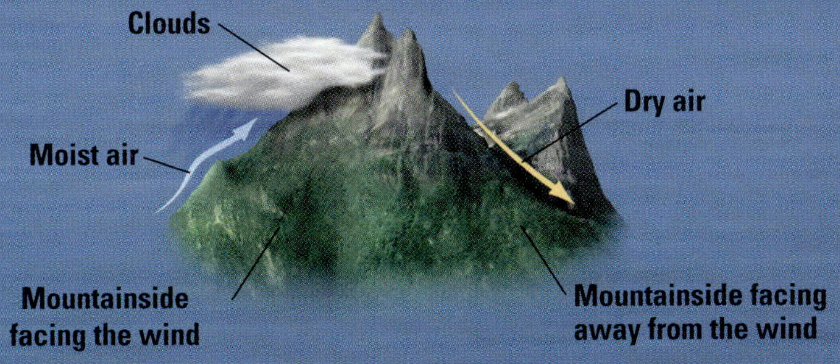


**What is El Niño?**

El Niño is a warm ocean current that visits the coasts of Chile and Peru about once every four to seven years, usually in the month of December. No one knows what causes El Niño to appear. This current is responsible for warming the waters off South America, killing marine life, and harming fishing industries. El Niño also has a bad effect on the weather. It can produce heavy rains and hurricanes. It may result in flooding in Florida and Louisiana, snowstorms in the Middle East, and a lack of rain in Australia and Indonesia. Lasting a couple of years, El Niño finally winds down, often to be replaced by a current that runs in the opposite direction—La Niña. In time, the weather returns to normal.

## One mountain, two climates

The valley situated on the western side of the Olympic Mountains in Washington state receives 150 inches (380 cm) of precipitation each year. The other side of the mountain, about 60 miles (100 km) away, receives less than 17 inches (43 cm)!

# Suffocating clouds

Humans have changed the world's climate. In cities, the asphalt and concrete used to construct roads and buildings hold in the Sun's heat. The climate of cities is not only warm, it can also be polluted. Factory chimneys and motor vehicles let off gases that trap heat close to Earth. The phenomenon of global warming worries many people because it has been getting steadily worse during the last 100 years. Over a long period of time, a temperature increase of just a few degrees worldwide is enough to cause polar ice to melt, which raises water levels in the ocean and can lead to the flooding of islands and low-lying coastal areas.

**Big headaches need big remedies**
Everyone must make an effort to save energy, avoid waste, and use low-pollution forms of transportation such as bicycles, electric cars, and carpooling. Even though the quality of our air has improved in the last 30 years, there is still much work to be done.

## Blinding smog

Big cities like Los Angeles and Mexico City are often covered by smog, a fog that occurs because of air pollution. In 1952, the city of London, England, was enveloped by smog so thick that people on the streets had to feel their way around by touching the walls of buildings!

# Earth: A giant greenhouse

Carbon dioxide, water vapor, ozone, sulfur dioxide, methane, CFCs (chlorofluorocarbons), and nitrous oxide form a belt of gases around Earth. This mixture of gases is responsible for a phenomenon called the greenhouse effect. Without this natural greenhouse effect, weather on our planet would be much cooler than it is now.

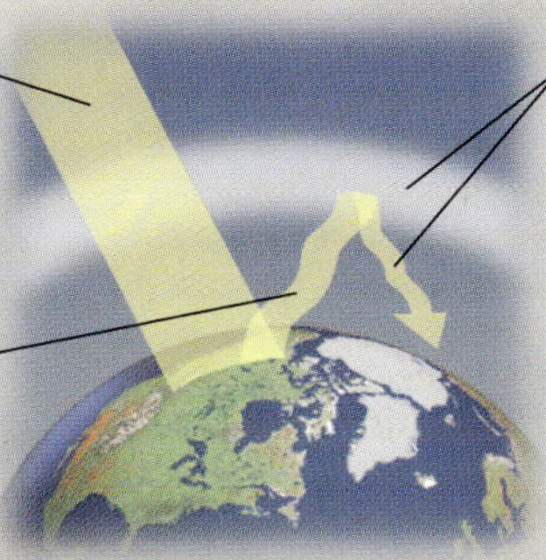


Human activities like farming and the cutting of trees, along with pollution, produce a large amount of greenhouse gases. Meteorologists fear that these gases dramatically add to Earth's natural greenhouse effect and increase global warming. This imbalance in the world's climate can cause flooding, hurricanes, and shortages of rain in some areas.

# Predictions . . . For better or worse

**Every day, meteorologists draw a picture of the current weather, then try to predict what it will do next. Because weather is always in motion, this is no easy job. Meteorologists use the most advanced technology to help them track the weather and make sense of it. After scanning the skies and studying their data, weather specialists come up with forecasts that help us plan for the hours and days to come.**

# Weather experts

Each morning, we may wonder what kind of weather the day will bring. By checking weather reports on the radio, television, or Internet, we can decide if it's better to go on a picnic or to a movie. Weather forecasts are particularly important to farmers, sailors, pilots, and others whose work is directly affected by weather conditions. People who forecast weather are called meteorologists. These experts not only tell us about the weather to come but also help keep us safe. Meteorologists use their knowledge to predict dangerous weather conditions. They can then alert entire communities to prepare for hurricanes and tornadoes that may be heading their way.

**The meteorologist: A multitalented scientist**
Many meteorologists have a college education in chemistry, physics, or mathematics. Some of them also need to have good communication skills so they can explain complicated information to the general public.

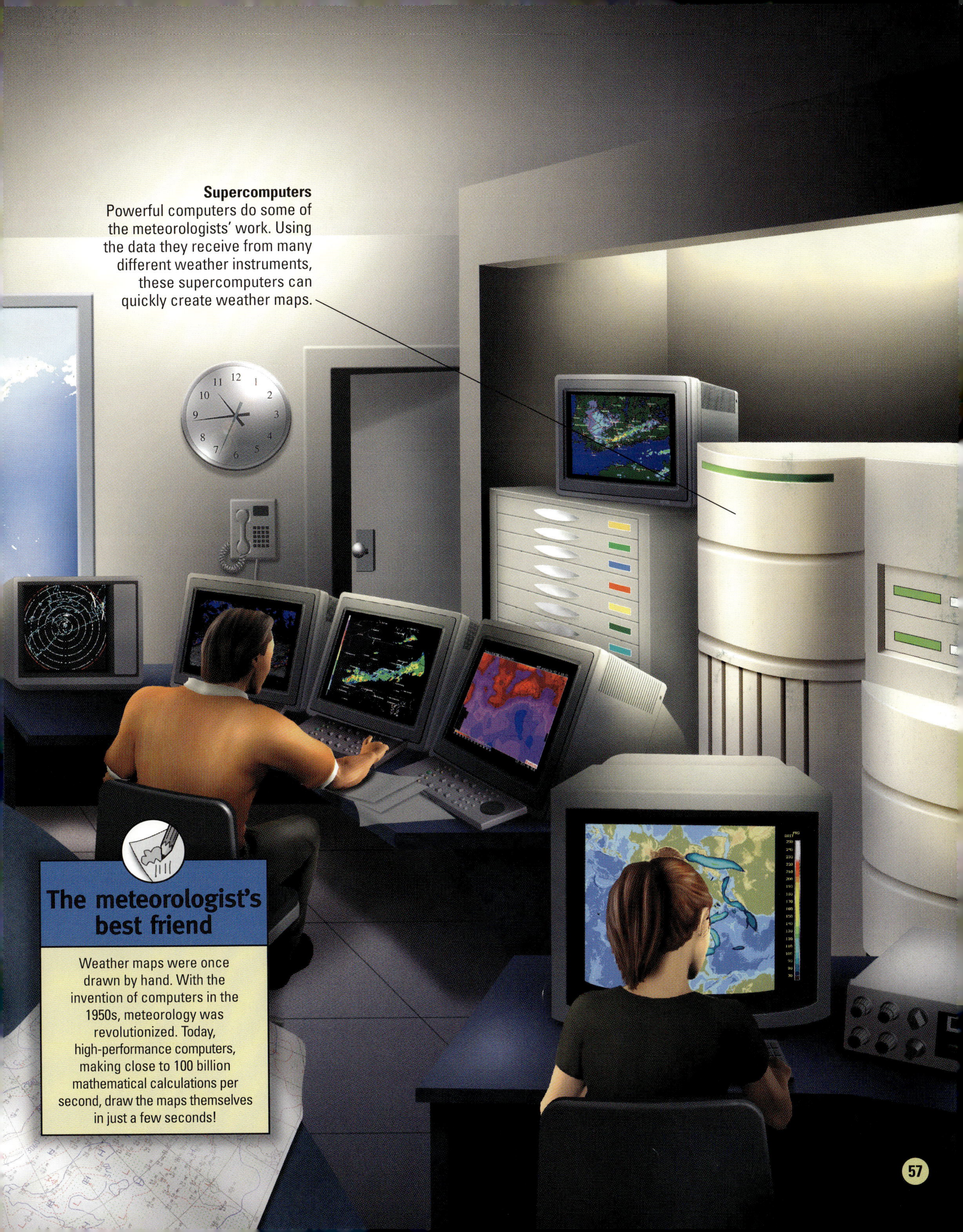


## The meteorologist's best friend

Weather maps were once drawn by hand. With the invention of computers in the 1950s, meteorology was revolutionized. Today, high-performance computers, making close to 100 billion mathematical calculations per second, draw the maps themselves in just a few seconds!

# Weather forecasting

Forecasting the weather in a particular area is not easy. Large amounts of weather data are needed. Several times a day, meteorologists receive reports from weather stations on land and from hundreds of ships, buoys, weather balloons, radar stations, and satellites. All this data about water vapor, temperature, precipitation, winds, clouds, pressure, and cloud cover is assembled and analyzed by powerful computers. Based on the data they receive, these computers create weather maps that group together many pieces of information. However, it still takes a person to forecast the weather for a particular city or region. Unlike most of today's machines, a meteorologist can take into account local factors that may affect the weather, such as the presence of hills or lakes.

## Four-legged weather forecasters

When dogs and cats don't move in their sleep, many people believe it is a sign that good weather will come. On the other hand, if cows are lying down in a field, frogs are croaking outside a pond, and birds are flying low, rain is on the way. Although nature can help us predict the weather, its signs should not be taken too seriously. After all, even animals can make mistakes!

# The secrets of weather maps

Every day, weather maps like this appear on television, in newspapers, and on the Internet. If you know the meanings of various symbols used on the maps, you will better understand the weather forecast.

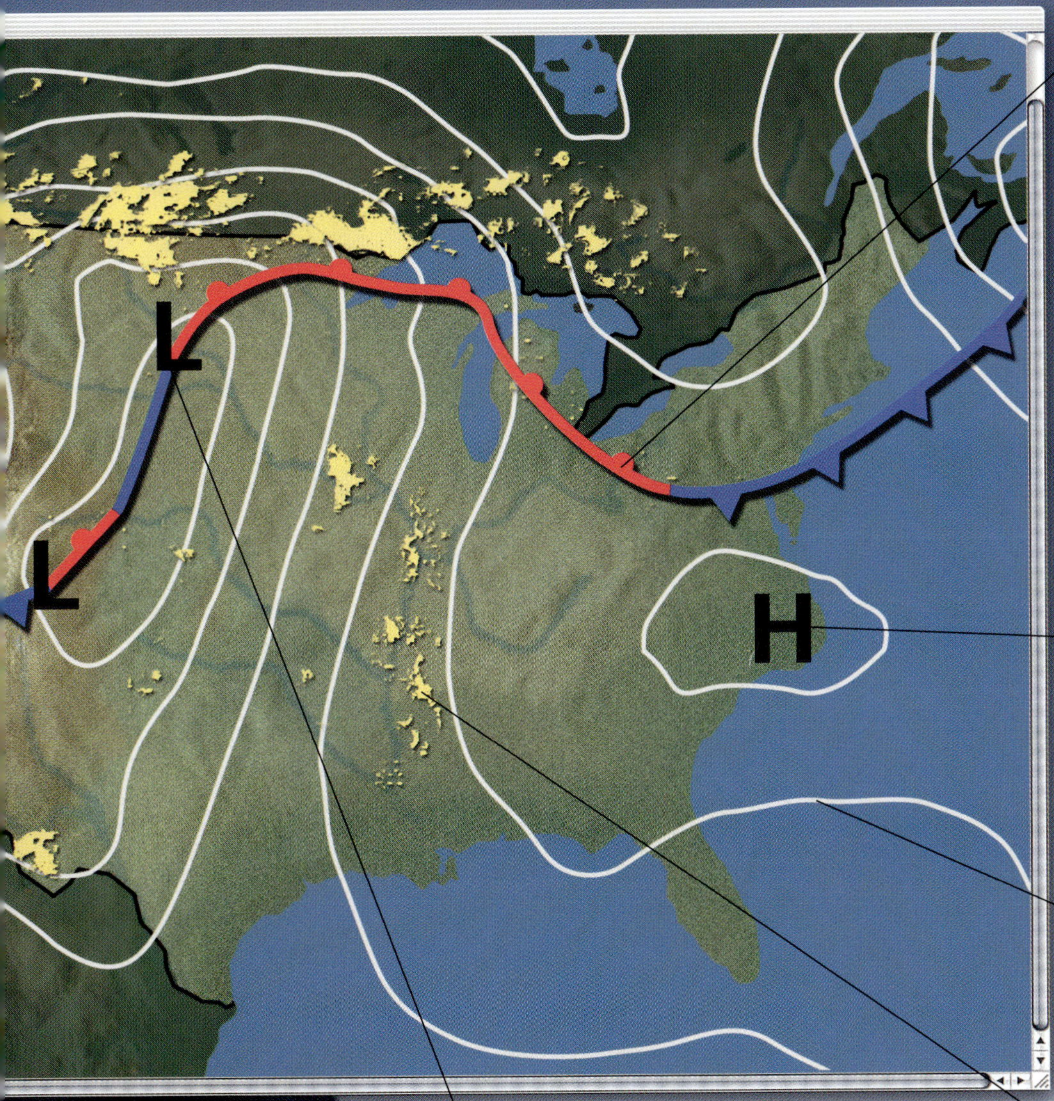


**Fronts**
A front is the border between a mass of warm air and a mass of cold air. The arrival of a front announces a change in the weather. On weather maps, fronts are represented by curved lines marked with triangles (for cold fronts) or semicircles (for warm fronts). The triangles and semicircles point in the direction the fronts are moving. A warm front causes temperatures to rise and often brings continuous rain or snow. On the other hand, a cold front brings colder temperatures, sudden showers, thunderstorms, and sometimes even hail.

**High-pressure zone**
A high-pressure (or anticyclone) zone is represented by the letter H. It shows an area where air pressure is high. A high-pressure area usually brings nice weather.

**Isobars**
Isobars are curved lines connected by points on the map that show where air pressure is the same.

**Low-pressure zone**
A low-pressure (or depression or cyclone) zone is represented by the letter L. It shows an area where the atmospheric pressure is low. Low air pressure can mean bad weather.

**Precipitation zones**
Areas with rain or snow are represented by yellow spots.

# Tools for measuring the weather

Every day, thousands of meteorologists in more than 170 countries collect data on wind speed, temperature, humidity, air pressure, hours of sunshine, and amount of rainfall or snowfall. To carry out their work, these weather experts use many different measuring instruments. Weather stations have a number of instruments grouped together in a common area. Several times a day, weather reports are produced by the instruments located in some 12,000 weather stations around the world.

**Wind vane**
The wind vane shows wind direction. The wind vane helps meteorologists to track, among other things, the movement of storm clouds.

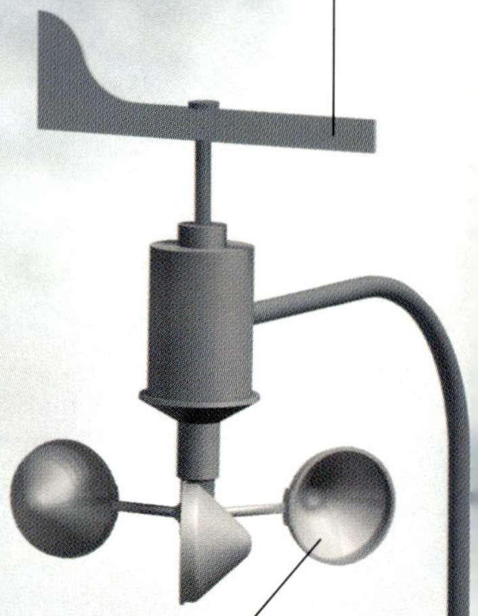

**Anemometer**
The anemometer measures wind speed. The harder the wind blows, the faster the cups spin around.

**Heliograph**
The heliograph is a glass ball that captures the Sun's rays and focuses their heat onto a piece of paper. The heat produces a scorch mark on the paper. The length of the scorch mark is measured to determine the hours of sunshine each day.

**Rain gauge**
The rain gauge is a container marked with small measurement lines. It shows the amount of rain that has fallen in an area.

**Snow gauge**
The snow gauge is used to collect snow. Meteorologists use it to calculate the amount of snow that has fallen in a particular area.

**Hygrograph**
The hygrograph measures and records water vapor or humidity in the air. Some hygrographs use human hairs to detect variations in the amount of water vapor. The hairs get longer in damp weather and shorter in dry weather.

**Automatic weather station**
Some places on Earth are difficult for meteorologists to reach. On the oceans, weather information is gathered by sailors and by weather stations attached to buoys. In the deserts, mountains, and polar regions, small weather stations function without the presence of humans. The data they collect is then sent by satellite to larger meteorological weather stations to be analyzed.

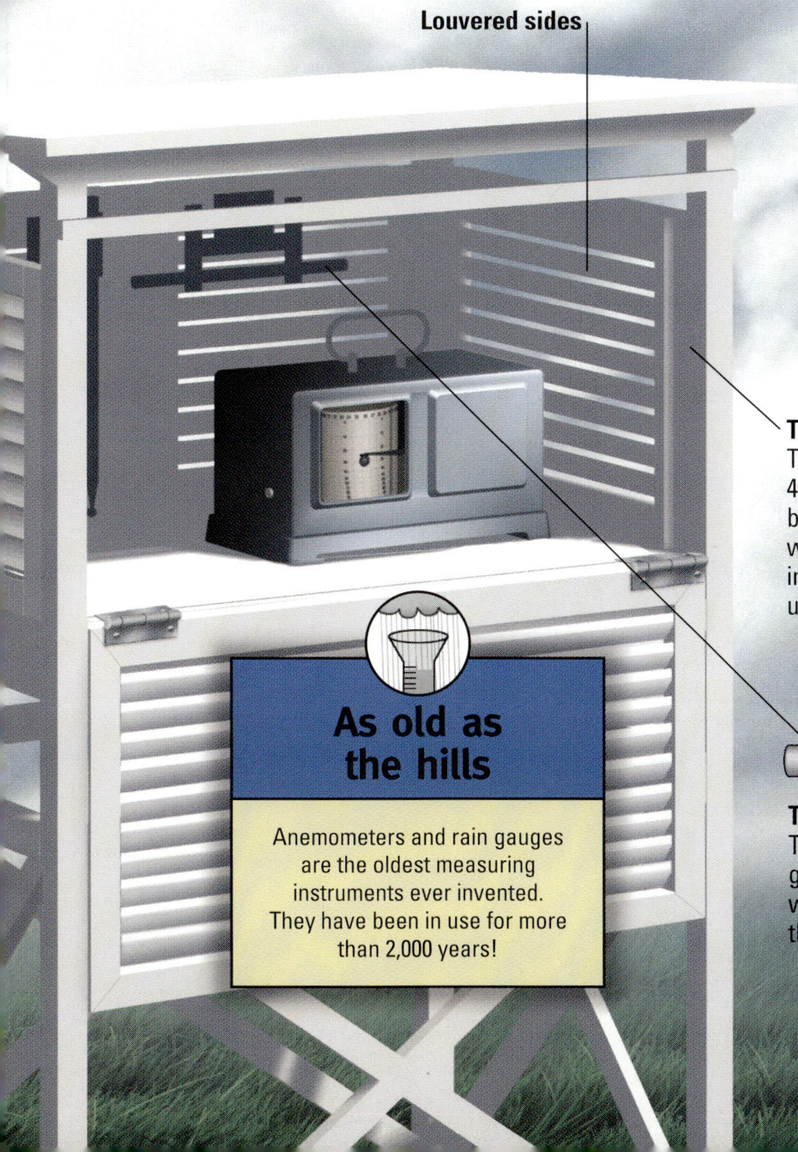

**Barometer**
The barometer measures air pressure. Variations in air pressure are a sign of changing weather. If the pressure drops rapidly, it signals bad weather to come. If the pressure rises, it means good weather is on the way.

See activity p. 70

**The Stevenson screen**
The Stevenson screen is a white-painted box that is mounted about 4 ft (1.2 m) above the ground. To prevent incorrect measurements, the box has louvered sides that allow for free airflow through the screen while preventing the Sun's rays from directly hitting the instruments inside. The Stevenson screen houses different weather devices used to measure temperature, pressure, and humidity of the air.

**Thermometer**
The thermometer is used to measure air temperature. It consists of a glass tube filled with a liquid, such as mercury or alcohol. As the air warms up, the liquid spreads and its level rises in the tube. When the air cools down, the liquid shrinks and its level drops.

## As old as the hills

Anemometers and rain gauges are the oldest measuring instruments ever invented. They have been in use for more than 2,000 years!

# Technology at the service of weather forecasting

In the last 50 years, modern technology has revolutionized meteorology. On the ground, powerful radars inspect the skies, searching for weather clues. Above the ground, more than 18 miles (30 km) up, weather balloons gather data on different air and cloud layers. Beyond Earth's atmosphere, satellites orbit in space, watching our planet from high above. These new ways of observing the weather have become very important to today's meteorologists. With the help of technology, they can follow the daily progress of the weather, predict where large storms are likely to occur, and study how Earth's climate is changing over the years.

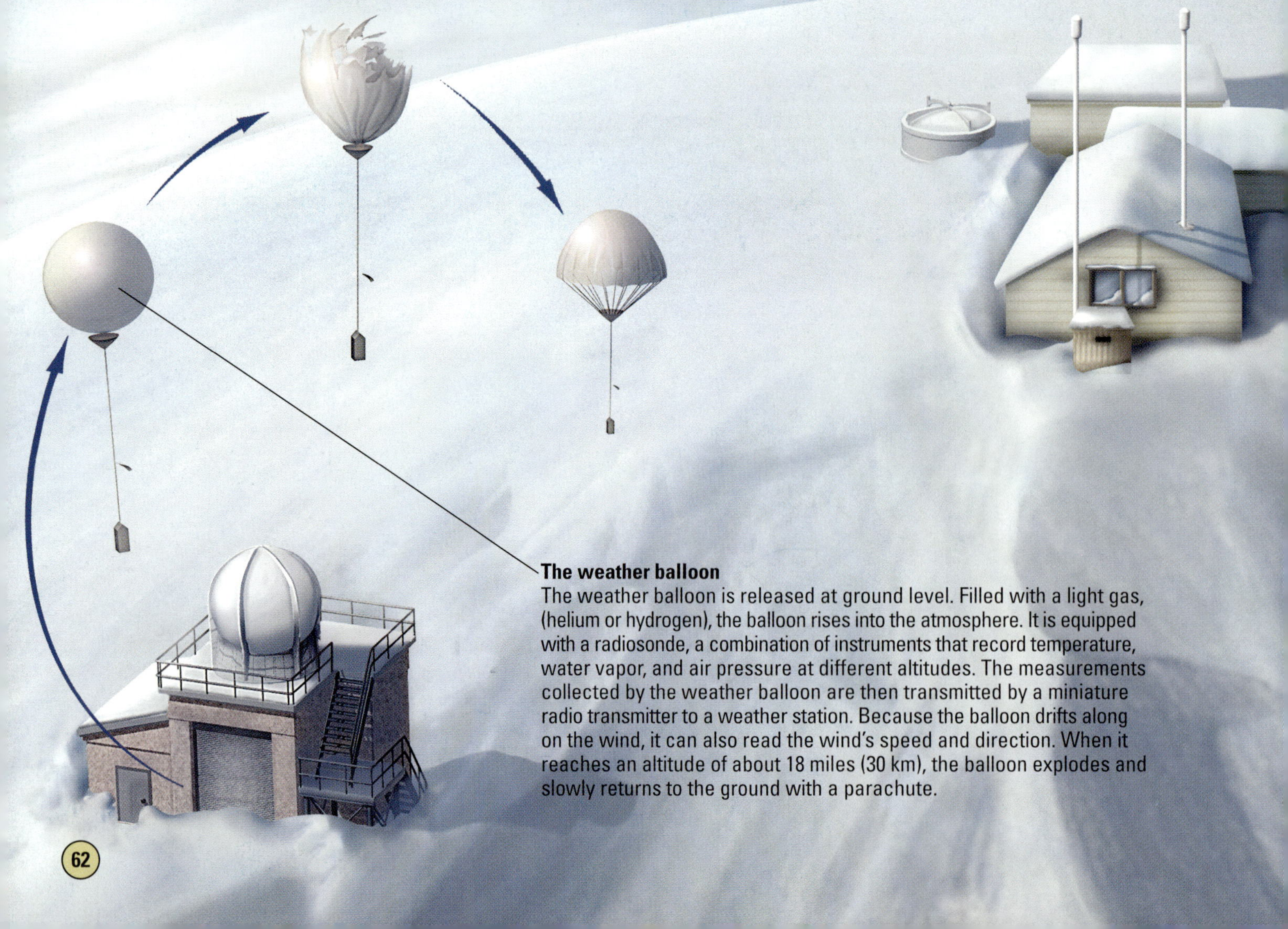

**The weather balloon**
The weather balloon is released at ground level. Filled with a light gas, (helium or hydrogen), the balloon rises into the atmosphere. It is equipped with a radiosonde, a combination of instruments that record temperature, water vapor, and air pressure at different altitudes. The measurements collected by the weather balloon are then transmitted by a miniature radio transmitter to a weather station. Because the balloon drifts along on the wind, it can also read the wind's speed and direction. When it reaches an altitude of about 18 miles (30 km), the balloon explodes and slowly returns to the ground with a parachute.

**Meteorological masts**
A meteorological mast is made up of several weather instruments attached to a pole at different heights.

**Satellites**
Weather satellites gather data that is transformed into images by computers. Newspapers and television often present satellite images that show cloud masses.
By comparing satellite images taken at different times, meteorologists can follow the movement of clouds, calculate wind speed, and predict precipitation. There are two types of satellites—geostationary and polar orbiting. Geostationary satellites surround the equator and always observe the same part of the planet. They can instantly gather atmospheric information from every part of the globe except the North and South poles. The polar orbiting satellites circle Earth, crossing the North and South poles. Orbiting at a low altitude, these satellites can precisely analyze the ground, the sea, and the lower atmosphere.

**Radar**
Located on the ground, radar helps meteorologists predict the type and amount of precipitation expected. Radar sends out radio waves that reflect back in different ways, depending on whether they hit raindrops, snowflakes, or hail. Using radar, meteorologists can also calculate wind speed and direction. This information is useful in tracking thunderstorms and tornadoes.

## A thousand eyes in the skies

Around the world, more than 1,000 weather balloons are released into the air twice a day. In addition, satellites carry out approximately 150,000 observations on a daily basis.

# Facts

## Climate

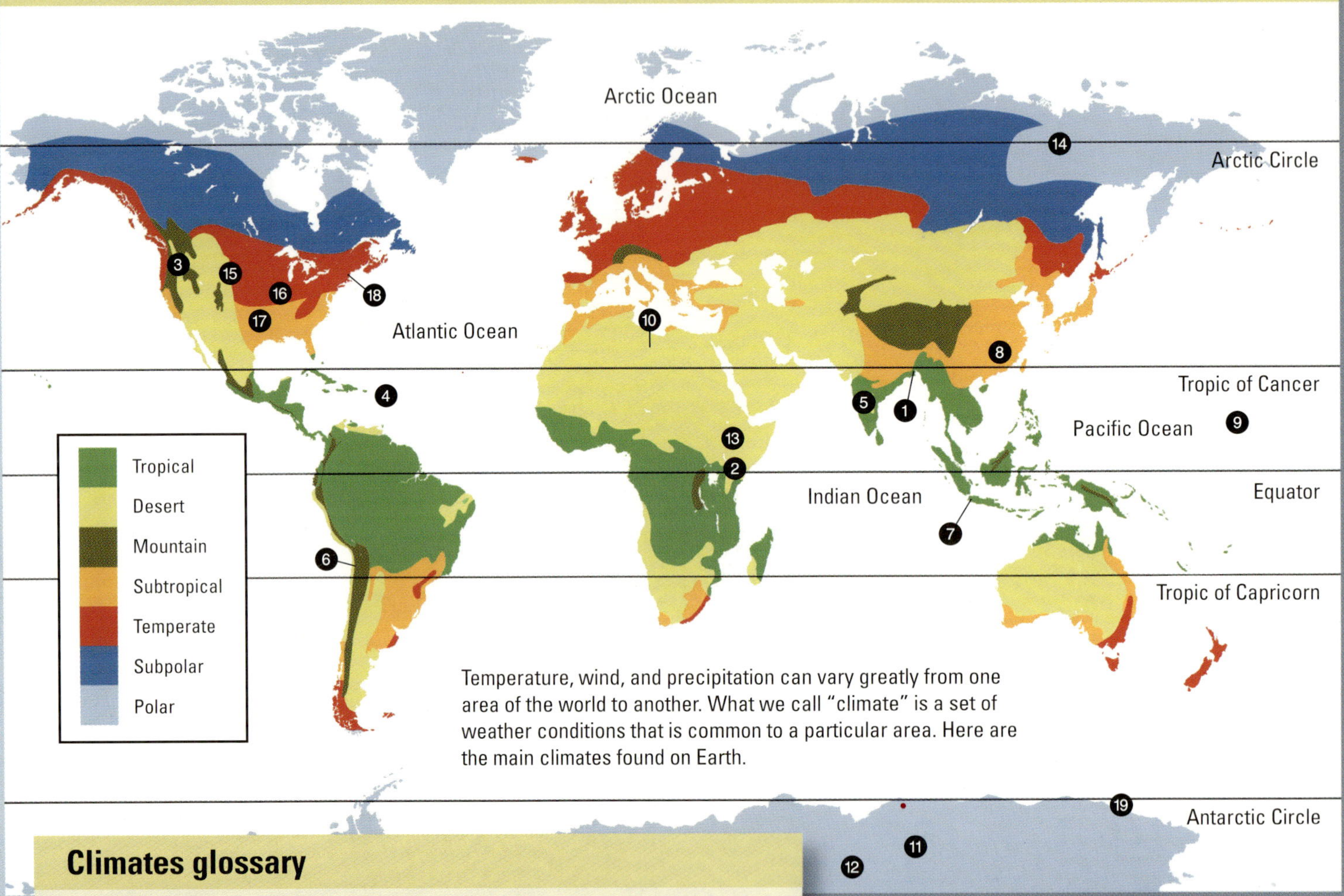


Temperature, wind, and precipitation can vary greatly from one area of the world to another. What we call "climate" is a set of weather conditions that is common to a particular area. Here are the main climates found on Earth.

### Climates glossary

**Tropical climate:** Very warm climate covering the area around the equator. There is little variation in temperature and much rainfall.

**Subtropical climate:** Climate marked by mild winters, hot summers, and much rainfall.

**Desert climate:** Climate marked by little rain or snow and almost no plant life. Temperatures vary greatly from daytime to nighttime.

**Mountain climate:** Climate found in mountainous areas. Temperatures drop and there is less plant life to be found as one climbs higher.

**Temperate climate:** Climate marked by variable weather and four very distinct seasons.

**Subpolar climate:** Climate marked by long, cold winters and short, cool summers.

**Polar climate:** Very cold climate where the ground is often frozen and where temperatures rarely rise above 50°F (10°C).

### The worst weather

If all of its atmospheric phenomena (cold, floods, lack of rain, heat, tornadoes, and tropical storms) are added up, the United States is the country with the worst weather. In just one year, the U.S.A. can expect to see 10,000 violent thunderstorms, 5,000 floods, 1,000 tornadoes, and several hurricanes.

## Record precipitation

The largest hailstone weighed 2.2 lb (1 kg). It was found in **Gopalganj** ❶, Bangladesh, on April 14, 1986.

The place where hail falls the most often is **Kericho** ❷, Kenya, with 132 days of hail a year.

From February 19, 1971, to February 18, 1972, more than 101 ft (31 m) of snow fell in **Paradise** ❸ on Mount Rainier in Washington state. That is the largest amount of snow ever to fall in an area in one year.

On November 26, 1970, in **Barst** ❹, Guadeloupe, more than 1.5 inches (3.8 cm) of rain fell in just 1 minute.

From August 1, 1860, to July 31, 1861, 86.81 ft (26.46 m) of rain fell in **Cherrapunji** ❺, India. That is the largest amount of rain ever to fall in one year.

The **Atacama** ❻ Desert in Chile is the driest place in the world. It has an average annual rainfall 4/1000 of an inch (0.1 mm)—just a few drops. In 1971, this region probably had its first rainfall in 400 years.

## Record storms

**Bogor** ❼, Indonesia, is the city that has been most often hit by thunderstorms. In 1916, thunderstorms struck on 322 days.

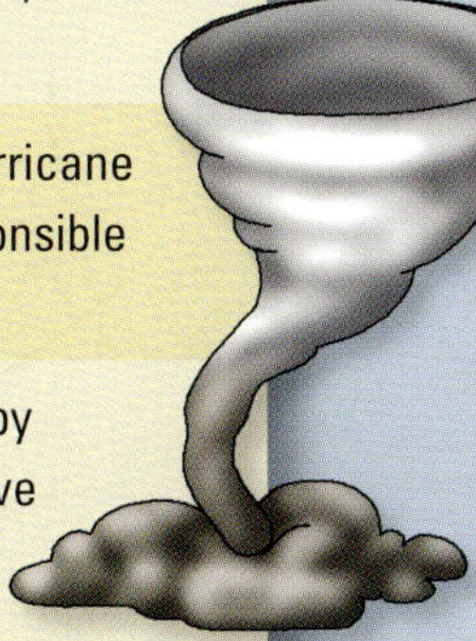

The deadliest and most destructive hurricane hit **Bangladesh** ❶ in 1970. It was responsible for the loss of at least 300,000 lives.

**China** ❽ is the country most often hit by floods. The biggest and most destructive one occurred in 1887 and caused the loss of more than 900,000 lives.

Cyclone John is the longest-lasting hurricane ever. During August and September 1994, it ran wild in the **Pacific Ocean** ❾ for about 30 days.

## Record temperatures

On September 13, 1922, the temperature in **Al Aziziyah** ❿, Libya, reached an all-time record high of 136°F (57.8°C). This temperature is hot enough to fry an egg!

On July 21, 1983, in **Vostok** ⓫, Antarctica, the temperature reached an all-time record low of -128.56°F (-89.2°C).

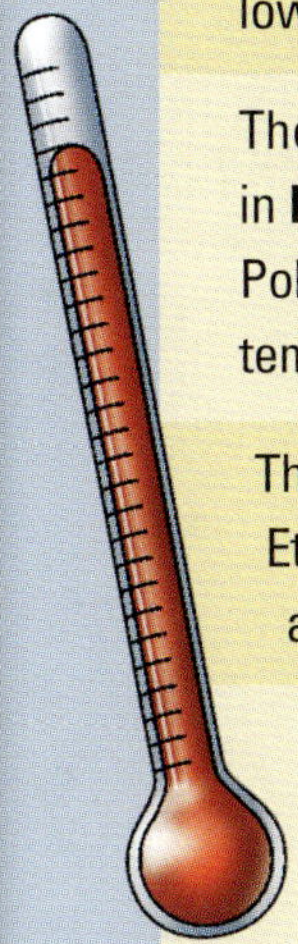

The coldest place in the world is Antarctica, in **Polus Nedostupnosti** ⓬, also called the Pole of Inaccessibility. Its average annual temperature is -72°F (-57.8°C).

The hottest place in the world is **Dallol** ⓭, Ethiopia. From 1960 to 1966, it had an average annual temperature of 93.9°F (34.4°C).

In **Verkhoïansk** ⓮, Siberia, extreme temperatures can range from -90°F (-68°C) in the winter to 90°F (32°C) in the summer. It is the biggest variation in temperature from one season to another.

In **Spearfish** ⓯, South Dakota, the temperature once soared from -4°F to 45°F (-20°C to 7°C) in just 2 minutes. That is the fastest temperature change ever.

## Record winds

The most destructive tornado in history hit **Missouri, Illinois, and Indiana** ⓰, on March 18, 1925. It traveled more than 215 miles (350 km) and was responsible for the loss of almost 700 lives.

The world's most powerful tornado hit **Oklahoma** ⓱ on May 3, 1999, with record winds of 316 mph (509 km/h).

On April 12, 1934, **Mount Washington** ⓲, in New Hampshire, recorded a gust of wind blowing at a speed of 231 mph (372 km/h).

The windiest place on Earth is **Commonwealth Bay** ⓳, Antarctica. Here, winds blow at an average annual speed of about 50 mph (80 km/h), and gusts of wind can reach 200 mph (320 km/h).

# Winds

Many kinds of winds blow in every part of the world. These winds can be gentle, warm, cold, dry, humid, or full of dust. Here are the names given to some of them:

**CHINOOK** 1
A warm wind from the west that blows in North America's Rocky Mountains. It can raise temperatures 40°F (22°C) in just 15 minutes, melting snow rapidly.

**CHOCOLATERO** 2
A warm and moderate north wind found in the area around the Gulf of Mexico. It is named after the chocolate-colored sand it blows into the air.

**BARBER** 3
This name is used in North America to describe a blizzard that blows from the north in the Gulf of St. Lawrence in winter. A powerful maritime storm, it brings drizzle and rain that instantly freezes everything, including hair and beards, which explains its name!

**MISTRAL** 4
A violent wind from the north that blows in central France almost year-round. Sometimes in the winter, the mistral is powerful enough to endanger trains traveling through the Rhône delta region.

**SIROCCO** 5
A warm, dry, dusty wind that blows mainly in the springtime from the Sahara desert northward to the Mediterranean Sea. As it crosses the sea, it turns humid and brings fog and rain to Malta, Sicily, and southern Italy.

**FOEHN** 6
A warm, dry wind from the south that blows mainly in spring and summer in the Valais region of Switzerland. Its powerful gusts blow downward from the mountains and bring fair weather and higher temperatures. It is so warm that it can melt snow faster than the Sun can!

**MONSOON** 7
A wind that changes direction with the seasons and occurs mainly in southern Asia. The summer monsoon blows from the southwest over the Indian Ocean toward Asia. This humid wind can cause extremely heavy rains to fall across India and surrounding areas. The winter monsoon is the opposite—a dry northeast wind that blows from the Asian continent toward the Indian Ocean.

**BURAN** 8
A strong northeast wind that blows across Siberia, other parts of Russia, and central Asia. It travels more than 35 mph (55 km/h), with violent gusts that often reduce visibility. In summer, it is called the black buran because of the dust it raises. In winter, it is usually called the white buran because of the snow it blows into the air. In Canada and the northern United States, this type of wind is called a blizzard.

# The Beaufort scale

Thanks to the Beaufort scale, we can calculate the force of the wind. This method of measurement was created in 1805 by a British sailor named Francis Beaufort. At that time, it was used to estimate the force of the wind without the use of any instruments. Admiral Beaufort developed his scale by observing the effects of winds on boat sails. Several years later, the Beaufort scale would also be used to measure the force of the wind on land.

| Force | 0 | 1 | 2 | 3 | 4 | 5 |
|---|---|---|---|---|---|---|
| Wind speed | 0–1 mph (less than 2 km/h) | 1–3 mph (2–5 km/h) | 4–7 mph (6–11 km/h) | 8–12 mph (12–19 km/h) | 13–18 mph (20–29 km/h) | 19–24 mph (30–39 km/h) |
| Description | Calm | Light air | Slight breeze | Gentle breeze | Moderate breeze | Fresh breeze |
| Effects | Chimney smoke rises straight up | Smoke indicates wind direction | Weather vanes move; wind felt on face | Small flags flutter; leaves and twigs move | Dust and paper rise; small branches move | Tree branches sway |

## Amazing skyscrapers

Palms and cycads are tropical trees that can easily survive natural disasters. During storms, they bend and sway, but are rarely knocked down. Builders were inspired by them to construct skyscrapers that can sway in the wind (for example, Sydney Tower in Australia). In high winds, the tops of some skyscrapers sway more than 3 ft (1 m)!

| 6 | 7 | 8 | 9 | 10 | 11 | 12 |
|---|---|---|---|---|---|---|
| 25–31 mph (40–50 km/h) | 32–38 mph (51–61 km/h) | 39–46 mph (62–74 km/h) | 47–54 mph (75–87 km/h) | 55–63 mph (88–101 km/h) | 64–73 mph (102–120 km/h) | More than 74 mph (+120 km/h) |
| Strong breeze | Near gale | Gale | Strong gale | Storm | Violent storm | Hurricane |
| Umbrella difficult to control; large tree branches wave; the wind whistles | Walking becomes difficult; trees sway | Small tree branches break off | Roof tiles fly off; tree branches break off | Major damage to houses; trees uprooted (rarely experienced) | Severe damage to houses (very rarely experienced) | Houses destroyed; landscape devasted |

# Weather symbols

Meteorologists use maps with circles that show where weather stations are located. Around each circle are various symbols and numbers that represent the weather being observed at these stations. This is what each symbol means:

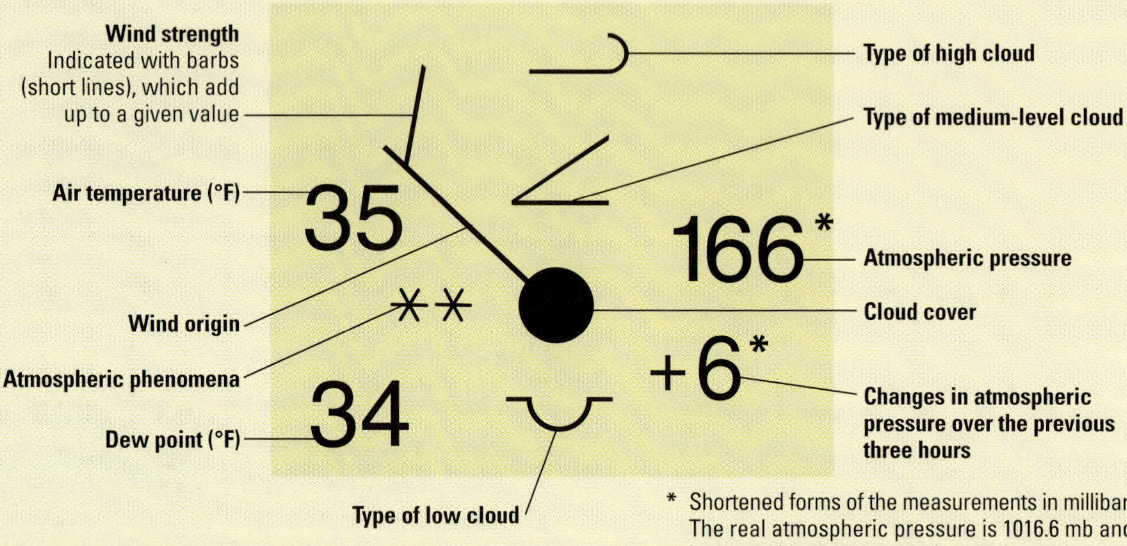


* Shortened forms of the measurements in millibars (mb). The real atmospheric pressure is 1016.6 mb and has increased by 0.6 mb over the previous 3 hours.

| Atmospheric phenomena |
|---|
| drizzle |
| light rain |
| moderate rain |
| heavy rain |
| light snow |
| moderate snow |
| heavy snow |
| hail |
| freezing rain |
| ice pellets |
| fog |
| thunderstorm |
| tornado |
| hurricane |

| Type of high clouds |
|---|
| cirrus |
| cirrocumulus |
| cirrostratus |

| Type of medium-level clouds |
|---|
| altostratus |
| altocumulus |

| Type of low clouds |
|---|
| stratus |
| stratocumulus |
| cumulus |
| cumulonimbus |
| nimbostratus |

| Cloud cover |
|---|
| clear sky |
| slightly covered sky |
| cloudy sky |
| very cloudy sky |
| overcast sky |

| Wind strength |
|---|
| still |
| half-barb 6 mph (10 km/h) |
| barb 12 mph (20 km/h) |
| barb + half-barb 18 mph (30 km/h) |
| pennant 58 mph (93 km/h) |

**Weather map**

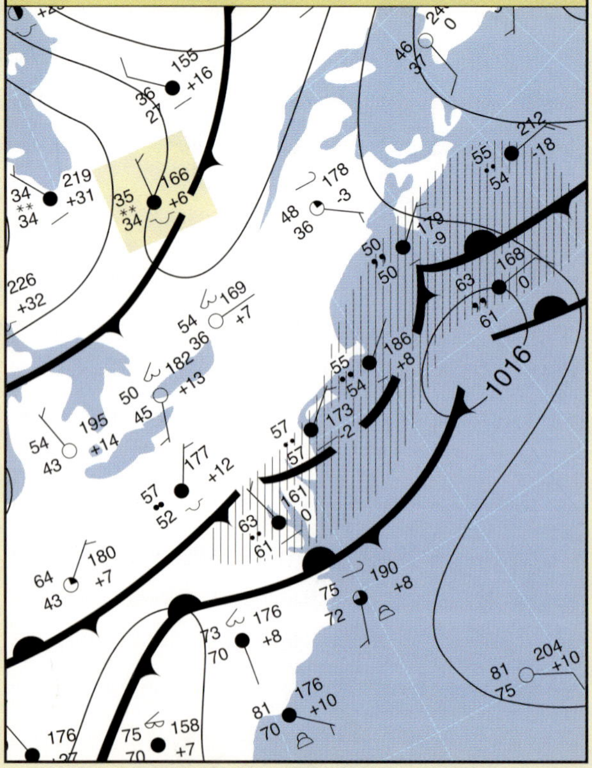

# Tricks for predicting the weather

Long before computers were helping meteorologists make forecasts, our ancestors were predicting the weather for the next day by observing signs in nature. By knowing the meaning of certain weather elements, you, too, can make predictions for the hours to come. Keep in mind that these elements are just signs and might not always be accurate. The more you observe, however, the more often your predictions are likely to come true.

## There will be fair weather in the next 24 hours if

- morning fog burns off a few hours after sunrise
- a light wind is blowing from the west (in the northern hemisphere) or from the east (in the southern hemisphere)
- clouds disappear in the late afternoon
- the smoke coming out of a chimney rises straight up
- there is dew or frost in the morning
- the moon is bright
- the setting or rising Sun looks like a ball of fire
- the sky is blue to the west
- the sky is red in the east at sunset

## There will be precipitation (rain or snow) in the next 24 hours if

- the clouds turn black in the afternoon in summer
- distant sounds are easier to hear
- odors from the ground (in marshes) are more present
- the wind increases or changes direction rapidly, switching from west to southwest to south and then to southeast
- a halo forms around the Sun or the Moon (the bigger the halo, the sooner the precipitation will come), especially in summer
- there is no dew on the ground in the morning
- a morning rainbow appears in the sky to the west
- the western sky is red at sunrise
- the setting Sun looks pale
- the clouds turn red at sunset
- birds and insects are flying low

## When lightning strikes, and strikes again

Roy C. Sullivan, a former forest ranger from Virginia, has been struck by lightning seven times! He has been knocked unconscious, lost a toenail, suffered injuries to his chest and stomach, and had his hair catch fire twice!

## It's raining frogs!

In 1946, it rained frogs in Memphis, Tennessee! This unusual phenomenon is often caused by a tornado. Sucking the amphibians out of their pond, the tornado carried them a distance, then spat them out of the sky like rain.

# Activities

## Make a water barometer

Atmospheric pressure, also called air pressure, corresponds to the weight of air. Measuring atmospheric pressure is very useful in predicting the weather. It is done using an instrument called a barometer. Here is how to make your own:

### Necessary materials

- a jar with a wide opening
- an empty bottle with a narrow neck
- water at room temperature
- food coloring
- a spoon
- a sticky label or a small piece of paper and adhesive tape
- a pencil
- a notebook

### Experiment

**1.** Pour the water into the jar until it is half full.

**2.** Add a few drops of food coloring and stir the water with the spoon until it is colored.

**3.** Put the bottle upside down in the jar so that the neck of the bottle is sitting about half an inch (1.25 cm) underwater.

**4.** Stick the label or the piece of paper to the side of the jar to help you measure the differences in the level of water in the bottle's neck.

**5.** Using your pencil, trace a line on the paper showing the water level in the bottle's neck.

**6.** Place your barometer away from heat and cold. Over a period of a few days, measure the level of water in the bottle's neck and write down if it rises or falls.

### Observe carefully

The water level in the bottle should change from one day to the next. When the air pressure is high, it pushes on the jar of water and makes the water rise in the neck of the bottle. High atmospheric pressure is usually a sign of good weather. On the contrary, when air pressure goes down, the level of water in the neck of the bottle goes down as well. Low atmospheric pressure is usually a sign of bad weather.

# Beating the heat

When the Sun is blazing hot, will you feel cooler if you wear dark-colored or light-colored clothing? Try the following experiment and you will have your answer.

## Necessary materials

- 2 identical glasses
- a sheet of white paper
- a sheet of black paper
- water at room temperature
- adhesive tape

## Experiment

1. Wrap the sheet of white paper around one glass and the sheet of black paper around the other glass. Hold the papers in place using the tape.

2. Fill both glasses with the room-temperature water.

3. Place both glasses in the sun for an hour, and then check the temperature of the water in each glass.

## Observe carefully

The water in the glass covered with the white paper has stayed cool, while the water in the other glass has heated up. This happens because the Sun's rays act differently upon objects they hit, depending on the objects' colors. Dark colors "capture" or absorb sunlight and change it into heat. The black paper around the glass captured the Sun's heat and then passed the heat on to the water inside. Light colors work in the opposite way. Similar to a mirror, light colors have the ability to partly reflect or bounce the Sun's rays. The white paper around the glass reflected the Sun's rays away from the glass and prevented the Sun's heat from reaching the water inside. Now you know how to dress during the next heat wave: to stay cooler, wear light colors like white, pink, or yellow, instead of dark colors like black, brown, or navy blue.

# Dress for the weather

When it's cold outside, it is better to cover up, especially your head! After doing this experiment, you will understand the wisdom of wearing a hat or cap in the wintertime!

## Necessary materials

- 2 identical jars
- a measuring cup
- hot water
- a cap

## Experiment

1. Using the measuring cup, fill the two jars with equal amounts of hot water. Be careful not to burn yourself!

2. Cover one of the jars with the cap.

3. Place both jars in the refrigerator and let them sit for 30 minutes.

4. Dip a finger into each bowl and compare the water temperatures.

## Observe carefully

The water in the jar covered with the cap is warmer. The reason is simple: the cap helps prevent heat from leaving the jar, so the water inside doesn't get cold as quickly. You can see why it's better to cover up in cold weather!

# Measure the weight of air

Although it is hard to imagine, air, like everything else, has weight. Try this experiment and find out for yourself.

### Necessary materials

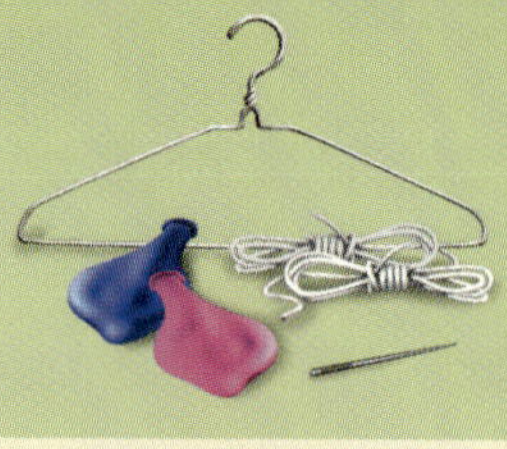

- a clothes hanger
- 2 balloons of equal size and shape
- a needle
- 2 pieces of string of equal length

### Experiment

**1.** Blow up the balloons. To ensure you make them the same size, count the number of times you exhale and try to blow as evenly as possible.

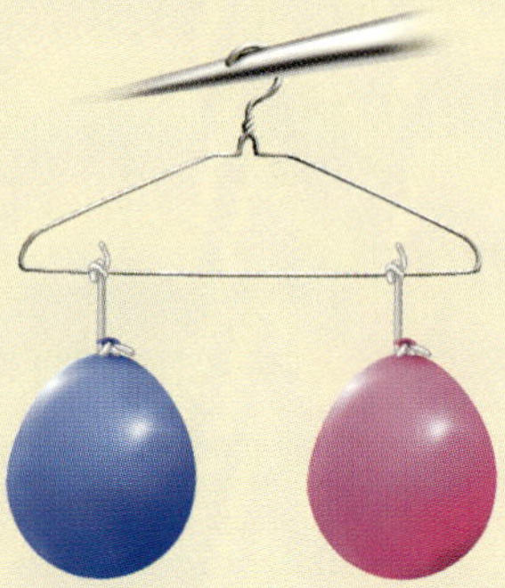

**2.** Using the strings, hang a balloon at each end of the hanger. Suspend the hanger.

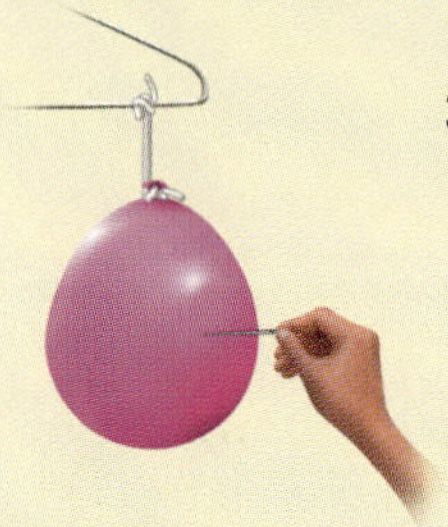

**3.** Once the hanger is balanced, use the needle to pop one of the balloons. Be careful not to hurt yourself, especially when the balloon bursts.

### Observe carefully

The balloon full of air is heavier and makes the hanger tip lower on one side. This is proof that air does indeed have weight! In meteorology, the weight of air is an important part of predicting the weather. When light air is circulating over an area, it creates a high-pressure, or "anticyclone," zone. This condition brings good weather. On the other hand, heavy air circulating over an area creates a low-pressure, or "depression," zone. Here, the weather will be cloudy with precipitation.

# Make a rainbow

Follow these few steps to create your own rainbow!

### Necessary materials

- a transparent glass
- water
- a sheet of white paper
- some sunlight

### Experiment

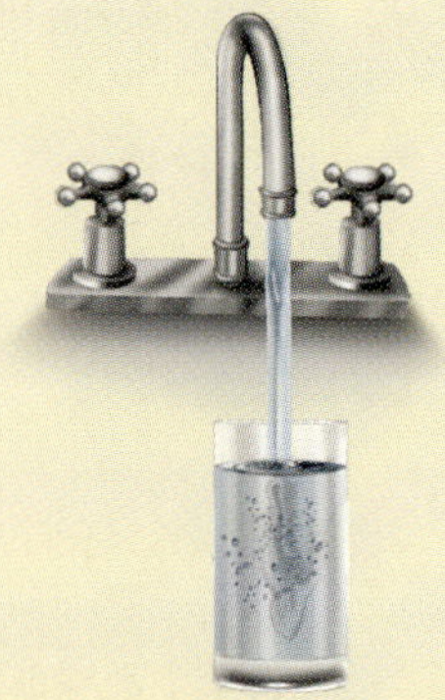

**1.** Fill the glass with water.

**2.** Hold up the glass to the sunlight. Hold up the sheet of paper directly under the glass. Make sure that the glass is in the sunlight but the paper is in the shade.

### Observe carefully

The band of colors that appears on your sheet of paper is the result of sunlight breaking down. Even though it looks white to us, sunlight is actually made up of several different colors. In this experiment, the water breaks down the Sun's white light and separates it into seven colors: red, orange, yellow, green, blue, indigo, and violet. This phenomenon is responsible for the rainbow that forms after a rainstorm when the Sun appears. Each raindrop acts the same way as the water in your glass. Together, the raindrops break down the Sun's white light and turn it into a fantastic rainbow.

# The effect of cold and heat on air

Try this experiment and you will understand why warm air rises while cold air stays at ground level.

## Necessary materials

- a plastic bottle with a narrow neck
- hot water
- cold water and ice cubes
- two dishes with raised edges (such as cake pans)
- a balloon
- a string or a measuring tape

## Experiment

**1.** Carefully pour the hot water into one of the dishes.

**2.** Pour the cold water into the second dish and add the ice cubes.

**3.** Blow up the balloon slightly then slip the open end over the neck of the plastic bottle.

**4.** Place the bottle in the dish of hot water for about 5 minutes. Hold the bottle securely to keep it from tipping over during the experiment.

**5.** Using the string or the measuring tape, measure the balloon around its middle.

**6.** Place the bottle into the cold water and leave it there for about 5 minutes.

**7.** Measure the balloon again.

## Observe carefully

The balloon is bigger when the bottle is sitting in the hot water. Here is why: The bottle in the hot water, as well as the air inside it, warms up. As the air warms, it takes up more space. We say that the air expands. The air becomes lighter and rises to the top of the bottle. It continues to move up into the balloon, which makes the balloon inflate even more. By placing the bottle in the cold water, the balloon shrinks. The bottle, as well as the air it contains, cools down. As the air cools, it takes up less space. We say that the air contracts. The air leaves the balloon and sinks back down into the bottle.

The phenomenon produced in your experiment also occurs in nature. Currents of warm air rise, while colder air, which is heavier, tends to stay at ground level. It is this movement of air that creates wind!

# Be a rainmaker

All the water on Earth is constantly in motion. Try this experiment and see for yourself!

## Necessary materials

- a large glass bowl
- a small round container (a yogurt container, for example)
- a facial tissue
- cotton balls
- glue
- a large rock
- small pebbles
- small flowers and bits of grass
- plastic food wrap
- a large elastic band
- water

## Experiment

**1.** Pour a little water into the large glass bowl.

**2.** Place some small pebbles in the center of the bowl.

**3.** Place the small plastic container on top of the pebbles. You can help to camouflage the sides of the container using more pebbles. Just make sure to leave the opening of the container free.

**4.** Decorate the pebbles with small flowers and bits of grass to recreate a natural-looking environment around the small container.

**5.** Cover the bowl with plastic food wrap and hold it in place with the elastic band. Place the rock on top of the plastic wrap. Let the weight of the rock stretch the wrap so that it hangs just above the small container.

**6.** You can glue cotton balls on a facial tissue and wrap the tissue around the rock to imitate a cloud. Put the glass bowl in the sun and leave it for a day.

## Observe carefully

At the end of the day you will see that the small container, which was empty at the start, now contains water! As impossible as this may seem, you have created rain. The heat of the Sun transformed the water in the large bowl into invisible vapor that rose inside the bowl. When the vapor encountered the plastic wrap, the vapor was transformed into water droplets that collected under the rock. Bit by bit, the droplets then fell into the small container, in much the same way that rain falls to Earth and fills lakes.

Just like the water in the large bowl, the water on the ocean's surface changes into vapor with the help of the Sun. The vapor rises in the sky and encounters cold air. The cold air acts like the plastic wrap and transforms the vapor into water droplets. Billions of these droplets in the sky collect to form clouds. The droplets in the clouds then fall back to Earth in the form of rain.

# Create your own clouds

Performing this experiment will make clouds appear right before your eyes!

## Necessary materials

- a transparent plastic bottle with a narrow neck
- an ice cube
- a sheet of dark-colored paper
- very hot water (the hotter the water, the easier it will be to see the clouds)

## Experiment

**1.** Fill the bottle with very hot water. Be careful not to burn yourself! Let it sit for 5 minutes to ensure that the bottle gets hot as well.

**2.** Empty about half the water out of the bottle.

**3.** Place the ice cube on the neck of the bottle. Place the sheet of dark-colored paper behind the bottle and watch closely.

## Observe carefully

The hot water in the bottle creates invisible vapor that rises to the top of the bottle. Here the vapor meets the cold air given off by the ice cube. After a few minutes in contact with the cold air, the water vapor cools and condenses to form the cloud you will see appearing on the bottle's inner surface. The same phenomenon is produced in the atmosphere when clouds are formed. When they become too big and too heavy to float, the water drops in the clouds fall in the form of rain.

# Collecting raindrops

In this experiment, you will be able to collect individual raindrops and observe the differences in them. To carry out this test, you will have to wait for a rainy day.

## Necessary materials

- a plastic bowl with a cover
- flour
- a spoon
- rain

## Experiment

**1.** Fill the bottom of the bowl with about half an inch (1.25 cm) of flour, then seal the bowl using the lid.

**2.** Go out in the rain. Don't forget your raincoat! Once you are outside, remove the cover from the bowl and let the rain fall on the flour for 5 to 10 seconds.

**3.** Put the cover back on the bowl and go indoors. Let the bowl sit for 20 minutes.

**4.** Open the cover and, using the spoon, carefully remove the raindrops that have hardened in the flour.

## Observe carefully

You have managed to capture separate raindrops. You can now have fun comparing the different sizes of drops. Try this experiment during a hard downpour and during a light shower. Compare the sizes of the drops you have collected. What do you notice? In general, the diameter (or measurement across) of a raindrop measures around 0.08 inches (2 mm). However, it can vary from 0.02 to 0.2 inches (0.5 to 5 mm). The harder the rainfall, the bigger the raindrops will be.

# Glossary

## A

**Altitude**
The height of something (a mountain or a cloud, for example) measured from sea level.

**Antarctic**
The ice-covered continent situated at the southern end of Earth, including the South Pole.

**Arctic**
The land and seas situated at the northern end of the Earth, including the North Pole.

**Astronomer**
A scientist who studies stars, planets, and other space objects as well as events that occur in space.

## B

**Boiling point**
The temperature at which a substance is hot enough to boil. Water boils at 212°F (100°C).

**Bush**
A dry area of land covered with shrubs, brush, and sometimes trees.

## C

**Carpooling**
The transportation of several people in one vehicle to save fuel and reduce traffic.

**Climate**
A pattern of weather phenomena common to a particular region of the planet. This pattern is measured over a long period of time.

**Coastal region**
An area bordering the ocean.

**Current**
The movement of water, air, or electricity in a specific direction.

## D

**Data**
Information collected from research or by observation.

**Diameter**
The length of a straight line that passes through the center of a round object.

**Downward-moving wind**
The downward movement of air.

## E

**Electrical particle**
A positively or a negatively charged particle that produces electricity by attracting a particle with the opposite charge.

**Equator**
An imaginary line that circles Earth midway between the two poles. It divides the planet into the Northern Hemisphere and the Southern Hemisphere.

**Equinox**
One of two days in the year having 12 hours of daylight and 12 hours of darkness. The equinox marks the beginning of spring and of fall.

## F

**Freezing point**
The temperature at which a substance is cold enough to freeze. Water freezes at 32°F (0°C).

## G

**Global warming**
An increase in the average temperature on Earth from one year to the next.

**Grassland**
A large expanse of grass in a tropical area.

**Greenhouse gases**
Gases that trap heat near Earth. These gases include carbon dioxide ($CO_2$), nitrous oxide, and methane.

## H

**Hemisphere**
The northern or southern half of Earth. The two hemispheres are divided by the equator.

**Humidity**
The amount of water vapor in the air.

## I

**Ice crystal**
A particle of ice formed in a cloud when a water droplet freezes.

## L

**Landslide**
A mass of land that detaches from a hillside and collapses, often because of rain. Landslides can cause a lot of damage.

## M

**Meteorology**
The study of various weather phenomena.

## N

**Negative charge**
An electrical charge that contains more negatively charged particles than positively charged particles.

## O

**Orbit**
The path taken by a planet or an object in space around another planet or object in space.

## P

**Pole**
The poles represent either the northern or the southern end of an imaginary line or axis around which Earth seems to rotate.

**Positive charge**
An electrical charge that contains more positively charged particles than negatively charged particles.

**Precipitation**
Water in liquid or solid form that falls from clouds and lands on Earth.

## S

**Solstice**
The day in the year marked by the most hours of daylight (summer solstice) or the fewest hours of daylight (winter solstice).

**Supercool**
The state of a substance that remains in liquid form when below its freezing point.

## T

**Tropical rain forest**
A forested area close to the equator. Its climate is warm and rainy, and temperatures here never drop below 64°F (18°C). The rain forest is home to a wide variety of plants and animals.

**Tropics**
The area on Earth around the equator where it is warm or hot all year. The tropics extend about 1,600 miles (2,575 km) north of the equator to the Tropic of Cancer and 1,600 miles (2,575 km) south of the equator to the Tropic of Capricorn.

## U

**Upward-moving wind**
The upward movement of air.

## V

**Vegetation**
A group of plants growing in an area.

**Visibility**
The quality of air that allows one to see a greater or lesser distance.

## W

**Water droplet**
A very small drop of water.

**Water vapor**
Water in the form of invisible gas.

# Index

**Bold = Main entry**

## Bibliography


Ahrens, C. Donald. *Meteorology Today.* West Publishing Company, 1994.

*The Handy Weather Answer Book.* Visible Ink Press, 1997.

*Understanding Climate and the Environment.* Québec Amérique, 2001.

*Weather.* Discovery Books, 1999.

*Weather.* Reader's Digest Explores, 1997.

Williams, Jack. *USA Today, The Weather Book.* Vintage, 1997.


## Photo credits


Page 27, landslide: © Crealp/Research Center on Alpine Environment / www.crealp.ch

Page 36, people of the desert: photo © www.danheller.com

Page 37, Inuit: © Galen Rowell/CORBIS/Magma

Page 39, dust devil: © Inflow Images – Australia

Page 41, forest fire: © Carol Polich/jhstock.com

Page 52, pollution: © Ted Spiegel/CORBIS/Magma

Page 56, meteorologist: © Ève Christian

Page 61, automatic weather station: © Martin B. Withers; Frank Lane Picture Agency/CORBIS/Magma


## Web sites

http://www.usatoday.com/weather/basics/wworks0.htm

http://www.msc-smc.ec.gc.ca/

http://beta.weather.com/

http://www.education.noaa.gov/cweather.html

## Acknowledgments


Benoît Allaire

City of Montréal

Gilles Brien (Environment Canada)


## Measurements

Most measurements in this book are written in abbreviated (shortened) form. Below you will find a key that explains what these abbreviations mean.

| Key to abbreviations | | |
|---|---|---|
| mm | = | millimeter |
| cm | = | centimeter |
| m | = | meter |
| km | = | kilometer |
| sq km | = | square kilometer |
| km/h | = | kilometers per hour |
| in. | = | inch |
| ft | = | feet |
| yd | = | yard |
| mph | = | miles per hour |
| lb | = | pound |
| sec | = | second |

**Conversion chart**

| Metric | U.S. |
|---|---|
| 1 cm | 0.4 in. |
| 1 m | 3.28 ft |
| 1 km | 0.62 mile |
| 10 km | 6.21 miles |
| 100 km | 62.14 miles |